SEMA SIMAI Springer Series

ICIAM 2019 SEMA SIMAI Springer Series

Volume 4

This sub-series of the SEMA SIMAI Springer Series aims to publish some of the most relevant results presented at the ICIAM 2019 conference held in Valencia in July 2019.

The sub-series is managed by an independent Editorial Board, and will include peer-reviewed content only, including the Invited Speakers volume as well as books resulting from mini-symposia and collateral workshops.

The series is aimed at providing useful reference material to academic and researchers at an international level.

More information about this subseries at http://www.springer.com/series/16499

Juan Pablo Muszkats • Silvia Alejandra Seminara •
María Inés Troparevsky

Editors

Applications of Wavelet Multiresolution Analysis

 Springer

Editors
Juan Pablo Muszkats
UNNOBA - Universidad Nacional del
Noroeste de la Provincia de Buenos Aires,
Buenos Aires, Argentina

Facultad de Ingeniería, Universidad de
Buenos Aires
Buenos Aires, Argentina

Silvia Alejandra Seminara
Facultad Regional Buenos Aires,
Universidad Tecnológica Nacional,
Buenos Aires, Argentina

Facultad de Ingeniería, Universidad de
Buenos Aires
Buenos Aires, Argentina

María Inés Troparevsky
Facultad de Ingeniería, Universidad de
Buenos Aires
Buenos Aires, Argentina

ISSN 2199-3041   ISSN 2199-305X   (electronic)
SEMA SIMAI Springer Series
ISSN 2662-7183   ISSN 2662-7191   (electronic)
ICIAM 2019 SEMA SIMAI Springer Series
ISBN 978-3-030-61715-8   ISBN 978-3-030-61713-4   (eBook)
https://doi.org/10.1007/978-3-030-61713-4

This Springer imprint is published by the registered company Springer Nature Switzerland AG.
The registered company address is: Gewerbestrasse 11, 6330 Cham, Switzerland

Preface

Signal analysis is a very wide set of techniques whose purpose is to obtain information from data, usually in the form of time series. The conjunction of signal analysis and modeling is the usual way through which it is possible to find out the underlying processes of a given phenomenon. In this context, preservation and suitable recovery of the frequency content of a signal are important problems to be faced. Therefore, it is essential to be able to separate the contributions of different frequency bands without losing resolution in the time domain. Wavelets are functions with unique properties that make them appropriate for analysis and hierarchical decomposition of signals.

The development of the theory of wavelets was led by Yves Meyer. In his publication *Ondelettes et Opérateurs* (Hermann, Paris, 1990), based on works by A.P. Calderón, A. Grossmann, J. Morlet, R.R. Coifman, and G. Weiss, Meyer founded a new chapter in harmonic analysis. Thenceforth, wavelet analysis has grown fast and incessantly, producing not only mathematical developments but also application-oriented research in many disciplines of science and engineering. In particular, since the publication of the famous article "Cycle octave and related transforms in seismic signal analysis" by P. Goupillaud, A. Grossmann, and J. Morlet (Geoexploration, 23, 1984), wavelet theory has been intensely developed, generating a great interest in diverse disciplines. As S. Mallat said, wavelets are based not on a "bright new idea," but on concepts that already existed under various forms in many different fields. The formalization and emergence of the wavelet theory is the result of a multidisciplinary effort that brought together mathematicians, physicist, and engineers, who recognized that they were independently developing similar ideas. During three decades of continuous growth, it has generated valuable results in both theoretical and applied areas where signal analysis is required.

The books *Wavelets: A Mathematical Tool for Signal Analysis* by C.K. Chui (SIAM, Philadelphia, 1997), *A Wavelet Tour of Signal Processing* by S. Mallat (Academic Press, San Diego, 1999), and *Ten Lectures on Wavelets* by I. Daubechies (SIAM, Philadelphia, 1992) provide a comprehensive presentation of the conceptual basis of wavelet analysis, including the construction and application of wavelet bases.

Why Wavelets?

Wavelets are oscillating functions, well localized in time and frequency domain, with zero mean and rapid decay in time. From one "mother wavelet," a two-parameter family of wavelets is obtained by translations in both the time variable and the scale factor (which influences the location of the time-frequency window and the width of the corresponding time and frequency windows). They constitute a basis that is usually but not necessarily orthogonal. These wavelet bases allow a function to be described in terms of a coarse overall shape, plus details, providing an elegant technique for representing the different scale components.

This localized transform in time and frequency can be successfully used to extract information from a signal that classical Fourier transform will not display. While the Fourier transform results in a representation that depends only on frequency, the windowed Fourier transform (or short-time Fourier transform) has a constant resolution with rigid time-frequency windows. They are poorly suited for signal analysis with a frequency spectrum that includes low and high frequencies.

On the other hand, the wavelet transform generates time-frequency windows with constant area, which are automatically adapted to provide a good time resolution for high frequencies (narrow time window) and good frequency resolution for low frequencies (wide time windows). In addition, the reconstruction method of the original function from its wavelet representation satisfies the valuable requirement of being stable under small perturbations and capable of giving arbitrarily high precision.

Why Multiresolution Analysis?

Multiresolution analysis can be seen as a sort of "microscope" capable of observing a function anywhere in its domain with a given resolution. Wavelets have suitable properties to carry on this type of analysis defining appropriate subspaces from dilations and translations of the mother wavelet.

In particular, the discrete wavelet transform in a multiresolution analysis scheme plays the role of a filter bank and offers the possibility of time-frequency localization, making it attractive for signals and image processing and pattern recognition in different applications: medicine, quantum physics, data compression, radar, and resolution of differential and integral equations, among others.

Multiresolution analysis (MRA) allows the signal to be decomposed by a bank of perfect reconstruction filters.

There exists a scale function whose translated versions generate the basis of a V_0 space, where it is accepted that the signal to be analyzed is included. The scaled versions of this scale function are used as the basis of the nested set of subspaces V_j (scale subspaces), where the lowest frequency part of the signal spectrum is found (the lower the farther from 0 is the integer "j").

The wavelet subspace W_j is the complement of V_j relative to V_{j-1} or V_{j+1}, depending on whether the MRA has been implemented using the convention of "j > 0" or "j < 0," respectively.

The j-th component of the signal is obtained by projecting it in the wavelet spaces W_j, whose basis is formed by the translations of a scaled version of the mother wavelet, and its frequency composition is mainly concentrated in the j-th frequency band in which the spectrum of the signal is broken up. The consecutive band centers are separated by an octave and the discretization is called dyadic.

The choice of the wavelet depends on several factors including the desired order of numerical accuracy and the computational effort.

Multiresolution analysis techniques are excellently developed in *An Introduction to Wavelet Analysis* (Applied and Numerical Harmonic Analysis) by D. F. Walnut (Birkhäuser, Boston, 2002). A clear and comprehensive exposition of the subject can be found in *An Introduction to Wavelets* by C.K. Chui (Academic Press, San Diego, 1992).

Since the mid-1990s and due to their desirable properties, researchers have paid attention to wavelet analysis in solving differential and integral equations. The solutions are approximated by wavelet and scaling expansions, with the advantage that multi-scale and localization properties can be exploited. It provides a robust and accurate alternative to traditional methods, specially when describing problems that have localized singular behavior.

In this book, we have gathered some works, presented in the mini symposium "Applications of Multiresolution Analysis with Wavelets," presented at the ICIAM 19, the International Congress on Industrial and Applied Mathematics, held at Valencia, Spain, in July 2019. The presented developments and applications cover different areas including filtering, signal analysis for damage detection, time series analysis, solution to boundary value problems, and fractional calculus. This bunch of examples highlights the importance of multiresolution analysis to face problems in several and varied disciplines.

Buenos Aires, Argentina   María Inés Troparevsky
May 2020

Contents

Editors and Contributors

About the Editors

Juan Pablo Muszkats has obtained a master's degree in mathematical engineering from the University of Buenos Aires. He presently works at the UNNOBA and Buenos Aires Universities, where he is professor of mathematical analysis and mathematical modelling, respectively. Professor Muszkats is an active member of the Non-Stationary and Non-Linear Time Series Analysis Group, where wavelet multiresolution analysis is applied to several phenomena.

Silvia Seminara is professor of mathematical analysis in the Faculty of Engineering, University of Buenos Aires. She obtained a master's degree in mathematical engineering at the University of Buenos Aires. Professor Seminara is the author of several scientific publications in areas of applied mathematics and mathematical education.

María Inés Troparevsky graduated from the University of Buenos Aires where she also obtained her PhD in mathematics. Currently, she is a professor in the Faculty of Engineering, University of Buenos Aires. She is the director of the research group Inverse Problems and Applications in the Faculty of Engineering where theoretical analysis and development of numerical algorithms to solve inverse problems are studied. Dr. Troparevsky is the author of several scientific publications. Her interests cover inverse problems, fractional calculus, and its applications.

Contributors

María B. Arouxet Facultad de Ciencias Exactas-CEMALP, Universidad de La Plata, La Plata, Argentina

Lucila Calderón Facultad de Ingeniería, Departamento de Ciencias Básicas, Universidad Nacional de La Plata, La Plata, Argentina

Marcela A. Fabio Centro de Matemática Aplicada, Universidad Nacional de San Martín, San Martín, Argentina

Guillermo La Mura Centro de Matemática Aplicada, Universidad Nacional de San Martín, Buenos Aires, Argentina

María T. Martín Facultad de Ingeniería, Departamento de Ciencias Básicas, Universidad Nacional de La Plata, La Plata, Argentina

Juan P. Muszkats Facultad de Ingeniería, Universidad de Buenos Aires, Buenos Aires, Argentina
Universidad Nacional del Noroeste de la Provincia de Buenos Aires, Buenos Aires, Argentina

Verónica E. Pastor Facultad de Ingeniería, Universidad de Buenos Aires, CABA, Argentina

Rosa Piotrkowski Escuela de Ciencia y Tecnología, Universidad Nacional de San Martín, Buenos Aires, Argentina
Facultad de Ingeniería, Universidad de Buenos Aires, Buenos Aires, Argentina

Miryam Sassano Universidad Nacional de Tres de Febrero, Buenos Aires, Argentina
Facultad de Ingeniería, Universidad de Buenos Aires, Buenos Aires, Argentina

Silvia A. Seminara Facultad de Ingeniería, Departamento de Matemática, Universidad de Buenos Aires, Buenos Aires, Argentina

Ricardo Sirne Facultad de Ingeniería, Departamento de Matemática, Universidad de Buenos Aires, Buenos Aires, Argentina

María Inés Troparevsky Facultad de Ingeniería, Departamento de Matemática, Universidad de Buenos Aires, Buenos Aires, Argentina

Victoria Vampa Facultad de Ingeniería, Departamento de Ciencias Básicas, Universidad Nacional de La Plata, La Plata, Argentina

Miguel E. Zitto Facultad de Ingeniería, Universidad de Buenos Aires, Buenos Aires, Argentina

Approximate Solutions to Fractional Boundary Value Problems by Wavelet Decomposition Methods

Marcela A. Fabio, Silvia A. Seminara, and María Inés Troparevsky

Abstract Fractional derivatives, unlike those of natural order, have "memory" and are useful to model systems where the past history is relevant. They are defined by means of integral operators, some of them having singular kernels, and calculations may be difficult. It is for that reason that it is necessary to develop numerical approximation methods to solve most of real problems. In this work we combine the wavelet transform with the fractional derivatives of a particular wavelet basis, by means of a Galerkin scheme, to build an approximate solution to boundary value problems involving Caputo-Fabrizio fractional derivatives. The numerical scheme is simple and stable, and its accuracy can be improved as much as desired. We present some numerical examples to show its performance.

1 Introduction

In the last decades, models described by fractional differential equations have appeared profusely in different areas of science. Several definitions of derivatives of non integer order have been proposed to fit different real phenomena requirements. A great quantity of results concerning solutions to this type of equations involving Riemann-Liouville, Caputo, Caputo-Fabrizio and Atangana-Baleanu fractional derivatives were stated [1–5], and different explicit and numerical solutions were developed [6–12].

Applications are numerous and in areas as varied as continuum mechanics [13], fluid convection and diffusion [14, 15], thermoelasticity [16], robotics [17], biology and medicine [18–21], computer viruses [22] or economics [23, 24].

M. A. Fabio
Centro de Matemática Aplicada, Universidad Nacional de San Martín, San Martín, Argentina
e-mail: mfabio@unsam.edu.ar

S. A. Seminara (✉) · M. I. Troparevsky
Facultad de Ingeniería, Departamento de Matemática, Universidad de Buenos Aires, Buenos Aires, Argentina

© The Author(s), under exclusive license to Springer Nature Switzerland AG 2021
J. P. Muszkats et al. (eds.), *Applications of Wavelet Multiresolution Analysis*,
SEMA SIMAI Springer Series 4, https://doi.org/10.1007/978-3-030-61713-4_1

In this work we adapt a methodology developed for fractional ordinary differential equations (FODE) to find approximate solutions to linear fractional partial differential equations (FPDE) involving Caputo or Caputo-Fabrizio fractional derivatives.

Succinctly, the idea of the proposed numerical scheme to solve linear FODE (see [25, 26]) consists of expressing the equation by means of the Fourier transform and decomposing the data and the unknown on a wavelet basis with appropriate properties: well localized in time and frequency domains, smooth, band limited and infinitely oscillating with fast decay. We project the data onto suitable wavelet subspaces and truncate it. Afterwards, through a Galerkin scheme, we calculate the coefficients of the unknown function in the chosen wavelet basis solving a linear system of algebraic equations that involves the fractional derivatives of the basis. Properties of the basis enable us to work on each level separately. Finally, we rebuild the solution from its wavelet coefficient. The proposed method is simple, since only the wavelet coefficients of the data and a matrix derived from the normal equations are needed. The error introduced in the approximation can be controlled improving the computation of the elements of the matrix and considering a more accurate truncated projection of the data. Properties of the basis and the operator guarantee that the resulting approximation scheme is efficient and numerically stable and no additional conditions need to be imposed. Details can be found in [25] and [26].

For the case of linear FPDE, we separate variables to obtain auxiliary FODE that we solve using the proposed scheme. We apply the methodology to solve a fractional diffusion equation with fractional time derivative. We compute the solution to a particular equation for different values of the fractional order of derivation $\beta \in (1, 2)$. When $\beta \to 2$, as expected, the behaviour of the solution is similar to that of the wave equation, which corresponds to the case $\beta = 2$.

This work is organized as follows: in the next section we present the fractional derivate operator; the wavelet basis and the approximation scheme are introduced in Sect. 3. In Sect. 4 a solution to the FPDE is proposed. Some numerical examples are presented in Sect. 5. Finally we state some conclusions.

2   Definitions and Properties

2.1   Caputo and Caputo-Fabrizio Fractional Derivatives

For $0 < \alpha < 1$ and f a function in $H^1((a, b))$, the Sobolev space of functions defined on (a, b) with $f' \in L^2((a, b))$, the Caputo fractional derivative (CFD) introduced in 1967 (see [27]) is defined as

$$
{}^C_a\mathcal{D}^\alpha_t f(t) := \frac{1}{\Gamma(1 - \alpha)} \int_a^t \frac{f'(s)}{(t - s)^\alpha} ds \tag{1}
$$

where Γ is the standard Gamma function and $-\infty \leq a < b$.

Caputo-Fabrizio fractional derivative (CFFD) introduced in 2015 (see [28]) is defined as

$$^{CF}_{a}\mathcal{D}^{\alpha}_{t} f(t) := \frac{M(\alpha)}{1-\alpha} \int_{a}^{t} f'(s) e^{-\frac{\alpha(t-s)}{1-\alpha}} ds \tag{2}$$

where $M(\alpha)$ is a normalizing factor verifying $M(0) = M(1) = 1$.

Both derivatives are integral operators that involve an integral from a to t, i.e. the past "history" of f is taken into account so, contrary to what happens in the natural order derivative case, they have "memory". It is worth noting that, in the case of Caputo-Fabrizio derivative, the integral operator has a regular kernel while in the Caputo case the kernel is singular.

Some properties of CFD and CFFD resemble those of classical derivatives: CFD and CFFD of order α of a constant function are zero and, for $0 < \alpha < 1$, $\lim_{\alpha \to 1} {}_{a}\mathcal{D}^{\alpha}_{t} f(t) = f'(t)$ and $\lim_{\alpha \to 0} {}_{a}\mathcal{D}^{\alpha}_{t} f(t) = f(t) - f(a)$.

Note that, when $a = -\infty$, both derivatives can be expressed as convolutions.

For the CFD, if κ is a causal function that coincides with $\frac{1}{t^{\alpha}}$ for $t > 0$, we have

$$^{C}_{-\infty}\mathcal{D}^{\alpha}_{t} f(t) = \frac{1}{2\pi \Gamma(1-\alpha)} \int_{\mathbb{R}} \widehat{f'}(\omega) \widehat{\kappa}(\omega) e^{i\omega t} d\omega, \tag{3}$$

where $\widehat{\kappa}(\omega) = \Gamma(1-\alpha)(i\omega)^{\alpha-1}$ and the circumflex represents the Fourier transform.

In the case of CFFD,

$$^{CF}_{-\infty}\mathcal{D}^{\alpha}_{t} f(t) = \frac{M(\alpha)}{1-\alpha} (f' * k)(t)$$

with the non-singular kernel $k(t) = e^{-\frac{\alpha t}{1-\alpha}}$, $t > 0$, from which

$$^{CF}_{-\infty}\mathcal{D}^{\alpha}_{t} f(t) = \frac{M(\alpha)}{2\pi(1-\alpha)} \int_{\mathbb{R}} \widehat{f'}(\omega) \widehat{k}(\omega) e^{i\omega t} d\omega \tag{4}$$

or

$$^{CF}_{-\infty}\mathcal{D}^{\alpha}_{t} f(t) = \frac{M(\alpha)}{1-\alpha} \int_{\mathbb{R}} \widehat{f}(\omega) m(\omega) e^{i\omega t} d\omega \tag{5}$$

with the not singular kernel $m(\omega) = \frac{1}{2\pi} \frac{i\omega}{\frac{\alpha}{1-\alpha}+i\omega}$. Expression (5) in terms of the Fourier transform will be considered in Sects. 3 and 4 to solve FPDE.

2.2   The Wavelet Basis

We briefly introduce the wavelet basis that we will use in the rest of the paper. Details and properties can be found in [29].

We recall that a wavelet is an oscillating function, well localized in time and frequency domains (see [30, 31]). For a special selection of the mother wavelet ψ, the family

$$\{\psi_{jk}(t) = 2^{j/2} \, \psi(2^j t - k), \; j, k \in \mathbb{Z}\}$$

is an orthonormal basis of the space $L^2(\mathbb{R})$ associated with a hierarchical structure of the space – the multiresolution analysis (MRA) – which is a sequence of nested subspaces V_j, the scale-subspaces, such that:

1. $V_j \subset V_{j+1}$;
2. $s(t) \in V_j$ if and only if $s(2t) \in V_{j+1}$;
3. if $s(t) \in V_0$ then $s(t+1) \in V_0$;
4. $\cup_{j \in \mathbb{Z}} V_j$ is dense in $L^2(\mathbb{R})$ and $\cap_{j \in \mathbb{Z}} V_j = \{0\}$;
5. there exists a function $\phi \in V_0$, called scaling function, such that the family $\{\phi(t - k), \; k \in \mathbb{Z}\}$ is an orthonormal basis of V_0.

The wavelet subspace $W_j = span\{\psi_{jk}(t), \; k \in \mathbb{Z}\}$ is the orthogonal complement of V_j in V_{j+1} and contains the detailed information needed to go from the approximation with resolution level j to the one corresponding to level $j + 1$:

$$\begin{cases} V_j \perp W_j \\ V_{j+1} = V_j \oplus W_j, \; j \in \mathbb{Z}. \end{cases}$$

Consequently

$$L^2(\mathbb{R}) = \bigoplus_{j \in \mathbb{Z}} W_j.$$

Moreover,

$$\begin{cases} V_n & = \bigoplus_{j<n} W_j \\ L^2(\mathbb{R}) & = [\bigoplus_{j \geq n} W_j] + V_n, \; \text{for any } n \in \mathbb{Z}. \end{cases}$$

The MRA is associated to an efficient method to compute the wavelet coefficients: the Mallat's algorithm (see [30]).

Looking for solutions to simple FODE as $^{CF}_{a}\mathcal{D}^{\alpha}_{t} f(t) = g(t)$ suggests the selection of the mother wavelet. Since the operators (3) and (4) act on Fourier transforms, it is convenient to implement a partition of the frequency domain in quasi-disjoint scale bands,

$$\mathbb{R}_\omega = \bigcup_{j=-\infty}^{\infty} \Omega_j ,$$

naturally associated with the wavelet subspaces W_j.

In order to achieve these benefits, we choose a Meyer wavelet: a band-limited function ψ, having a smooth Fourier transform $\widehat{\psi}$. In [29] we define the scale function and the wavelet as

$$\widehat{\phi}(\omega) = \begin{cases} 1 & |\omega| \leq \pi - \beta \\ \dfrac{v_\beta(\omega)}{\sqrt{v_\beta^2(\omega) + v_\beta^2(2\beta - \omega)}} & \pi - \beta < |\omega| < \pi + \beta \\ 0 & |\omega| \geq \pi + \beta \end{cases}$$

with

$$v_\beta(\omega) = \begin{cases} \exp\left(-\dfrac{(\frac{\omega-\pi+\beta}{2\beta})}{1 - (\frac{\omega-\pi+\beta}{2\beta})^2}\right) & |\omega - \pi + \beta| < 2\beta \\ 0 & |\omega - \pi + \beta| \geq 2\beta \end{cases}$$

and

$$\widehat{\psi}(\omega) = \sqrt{\phi^2(\omega/2) - \phi^2(\omega)} \; e^{-i\omega/2}$$

with parameter $0 < \beta \leq \pi/3$.

We recall that $\psi \in S$, the Schwartz class, and the family $\{\psi_{jk}, k \in \mathbb{Z}\}$ is an orthonormal basis of $L^2(\mathbb{R})$ associated to a MRA, well localized in both, time and frequency domain. Its spectrum, $|\widehat{\psi}(2^{-j}\omega)|$, is supported on the two-sided band

$$\Omega_j = \left\{ \omega : 2^j(\pi - \beta) \leq |\omega| \leq 2^{j+1}(\pi + \beta) \right\} \tag{6}$$

for some $0 < \beta \leq \pi/3$.

In Fig. 1 we show the graphs of ψ and $|\widehat{\psi}|$.

It is important to highlight that the sets $\Omega_{j-1}, \Omega_j, \Omega_{j+1}$ have little overlap (see Fig. 2) and W_j is nearly a basis for the set of functions whose Fourier transform has support in Ω_j. When solving FPDE this property will enable as to work on each level separately.

Details on the basis and its properties can be found in [29]. In [31] approximations of Sobolev, Besov and other functional spaces, using wavelets in the Schwartz class, are developed.

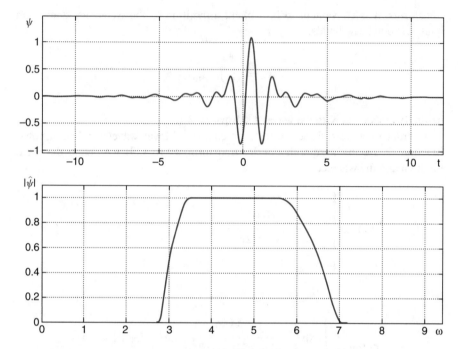

Fig. 1 Mother wavelet for $\beta = \pi/4$ (above) and $|\widehat{\psi}|$ for $\omega \geq 0$ (below)

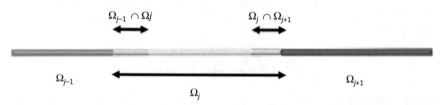

Fig. 2 The supports Ω_{j-1}, Ω_j, Ω_{j+1} and their overlapping

3 Approximate Solutions to a Fractional Initial Value Problem

In this section we resume the technique introduced in [32] to calculate approximate solutions to the following initial value problem:

$$\begin{cases} {}^{CF}_{a}\mathcal{D}^{\alpha}_t h(t) + \lambda_1 h'(t) + \lambda_0 h(t) = r(t) & \forall t \in (0, b) \\ \qquad\qquad\qquad h(0) = 0 \end{cases} \tag{7}$$

where $a \in \mathbb{R}$, $-\infty \leq a \leq 0$, h is the unknown, r is a known source function verifying $r(0) = 0$, and $\lambda_0, \lambda_1 \in \mathbb{R}$. Solution to this type of fractional initial value problem (FIVP) will be used to construct approximate solutions to FPDE by separating variables.

We will develop the calculations considering the CFFD. The case of CFD is similar. Details can be found in [25] for the case of CFFD and in [26] for the CFD.

First we will consider $a = -\infty$ in (7). Afterwards we will adapt the procedure to the case $a \neq -\infty$.

3.1   The Data

Recall that, for any $J \in \mathbb{Z}$, the data function $r \in L^2(\mathbb{R})$ can be decomposed as

$$r(t) = \sum_{j \in \mathbb{Z}} (\mathcal{Q}_j r)(t) = (\mathcal{P}_J r)(t) + \sum_{j \geq J} (\mathcal{Q}_j r)(t) =$$

$$= \sum_{n \in \mathbb{Z}} \langle r, \phi_{Jn} \rangle \, \phi_{Jn}(t) + \sum_{j \geq J} \sum_{k \in \mathbb{Z}} \langle r, \psi_{jk} \rangle \, \psi_{jk}(t)$$

where $\mathcal{Q}_j r$ and $\mathcal{P}_j r$ are the orthogonal projections of r in W_j and V_j, respectively.

We choose $J_{min}, J_{max} \in \mathbb{Z}$ so that the energy of r is concentrated in levels $J_{min} \leq j \leq J_{max}, k \in \mathbb{K}_j$:

$$r(t) \cong \sum_{j=J_{min}}^{J_{max}} r_j \cong \sum_{j=J_{min}}^{J_{max}} \tilde{r}_j(t)$$

where $r_j = \sum_{k \in \mathbb{Z}} c_{jk} \psi_{jk}$ is the projection of r on W_j, $c_{jk} = \langle r, \psi_{jk} \rangle$ are the wavelet coefficients, and $\tilde{r}_j$ is the truncated projection on W_j: $\tilde{r}_j = \sum_{k \in \mathbb{K}_j} c_{jk} \psi_{jk}$ for $\mathbb{K}_j \subset \mathbb{Z}$, $card(\mathbb{K}_j) = \eta_j < \infty$, satisfying

$$\sum_{k \notin \mathbb{K}_j} |\langle r, \psi_{jk} \rangle|^2 < \epsilon \|r_j\|_2^2$$

for certain ϵ near 0.

3.2   A Solution to the FDE

Let us decompose the solution of (7) in the basis, $h(t) = \sum_{j \in \mathbb{Z}} \sum_{k \in \mathbb{Z}} b_{jk} \psi_{jk}(t)$, and replace it in the equation:

$$\sum_{j \in \mathbb{Z}} \sum_{k \in \mathbb{Z}} b_{jk} \, [_{-\infty}^{CF} \mathcal{D}_t^\alpha \psi_{jk}(t) + \lambda_1 \psi'_{jk}(t) + \lambda_0 \psi_{jk}(t)] = \sum_{j \in \mathbb{Z}} r_j(t).$$

We note that

$$
{}^{CF}_{-\infty}\mathcal{D}^\alpha_t h(t) + \lambda_1 h'(t) + \lambda_0 h(t) =
$$

$$
= \tfrac{M(\alpha)}{1-\alpha} \int_{\mathbb{R}} \widehat{h}(\omega) m(\omega) e^{i\omega t} d\omega + \tfrac{\lambda_1}{2\pi} \int_{\mathbb{R}} i\omega \widehat{h}(\omega) e^{i\omega t} d\omega + \tfrac{\lambda_0}{2\pi} \int_{\mathbb{R}} \widehat{h}(\omega) e^{i\omega t} d\omega
$$

$$
= \int_{\mathbb{R}} \widehat{h}(\omega) H(\omega) e^{i\omega t} d\omega = r(t),
$$

where

$$
H(\omega) = \frac{-\lambda_1(1-\alpha)\omega^2 + i\left(M(\alpha) + \alpha\lambda_1 + (1-\alpha)\lambda_0\right)\omega + \alpha\lambda_0}{2\pi(\alpha + i\omega(1-\alpha))}. \tag{8}
$$

Thus, the images of the basis ψ_{jk} through the fractional differential operator $L(h) = {}^{CF}_{-\infty}\mathcal{D}^\alpha_t h + \lambda_1 h' + \lambda_0 h$ result in

$$
u_{jk}(t) = {}^{CF}_{-\infty}\mathcal{D}^\alpha_t \psi_{jk}(t) + \lambda_1 \psi'_{jk}(t) + \lambda_0 \psi_{jk}(t)
$$
$$
\tag{9}
$$
$$
= \int_{\mathbb{R}} \widehat{\psi}_{jk}(\omega) H(\omega) e^{i\omega t} d\omega.
$$

From (9), we note that the Fourier transform of u_{jk} satisfies $\mathrm{supp}(\widehat{u}_{jk}) \subset \Omega_j$. Then, based on previous observations, we can consider $u_{jk} \in W_j$ and, consequently, we can work on each level $J_{min} \le j \le J_{max}$ separately.

For a fixed j, we restrict ourselves to $\mathbb{K}_j$ to obtain

$$
{}^{CF}_{-\infty}\mathcal{D}^\alpha_t h_j(t) + \lambda_1 h'_j(t) + \lambda_0 h_j(t) = \sum_{k \in \mathbb{K}_j} b_{jk} u_{jk}(t) \cong \tilde{r}_j(t).
$$

The coefficients b_{jk} can be computed from the normal equations

$$
\sum_{k \in \mathbb{K}_j} b_{jk} \langle u_{jk}, \psi_{jm} \rangle = \sum_{k' \in \mathbb{K}_j} c_{jk'} \langle \psi_{jk'}, \psi_{jm} \rangle
$$

or, in matrix form,

$$
(\mathcal{M}^j \mathbf{b}^j)_k = \mathbf{c}^j_k, \quad k \in \mathbb{K}_j, \tag{10}
$$

where $\mathcal{M}^j_{km} = \langle u_{jk}, \psi_{jm} \rangle$.

We approximate $h_j(t)$ by $\tilde{h}_j(t) = \sum_{k \in \mathbb{K}_j} b_{jk} \psi_{jk}(t)$, with $(\mathbf{b}^j)_k = b_{jk}, k \in \mathbb{K}_j$ from (10), and $h(t) \cong \sum_{j=J_{min}}^{J_{max}} \tilde{h}_j(t)$.

Since r is causal, $r(0) = 0$, its wavelets coefficients c_{jk}, associated to $t \le 0$, are almost null and $\mathbf{b}^j_k$ results nearly null. Thus the proposed solution satisfies $h(0) = 0$.

We can proceed in a similar way if higher (natural) order derivatives appear in (7).

Further, we can adapt the scheme to the case where initial conditions are not null or $a \neq -\infty$, particularly, $a = 0$.

If $a = 0$, we consider $\overline{h} = h \cdot \chi_{[0,b]}$.

For $t < 0$ we have $\overline{h}'(t) = 0$ and $\overline{h}(t) = 0$ and, for $t > 0$, $\overline{h}'(t) = h'(t)$.

In addition,

$$
\begin{aligned}
{}^{CF}_{-\infty}\mathcal{D}^\alpha_t h(t) + \lambda_1 h'(t) + \lambda_0 h(t) &= e^{-\frac{\alpha t}{1-\alpha}} \left({}^{CF}_{-\infty}\mathcal{D}^\alpha_t h \right)(0) + {}^{CF}_0 \mathcal{D}^\alpha_t h(t) + \lambda_1 h'(t) + \lambda_0 h(t) \\
&= e^{-\frac{\alpha t}{1-\alpha}} [r(0) - \lambda_1 h'(0) - \lambda_0 h(0)] + {}^{CF}_0 \mathcal{D}^\alpha_t \overline{h}(t) + \lambda_1 \overline{h}'(t) + \lambda_0 \overline{h}(t) \\
&= {}^{CF}_0 \mathcal{D}^\alpha_t \overline{h}(t) + \lambda_1 \overline{h}'(t) + \lambda_0 \overline{h}(t),
\end{aligned}
$$

and $\overline{h}(0) = 0$.

Thus $\overline{h}$ satisfies

$$
{}^{CF}_0 \mathcal{D}^\alpha_t \overline{h}(t) + \lambda_1 \overline{h}'(t) + \lambda_0 \overline{h}(t) = {}^{CF}_{-\infty}\mathcal{D}^\alpha_t h(t) + \lambda_1 h'(t) + \lambda_0 h(t) = r(t), \quad t > 0.
$$

When initial conditions are not null ($h(0) = h_0 \neq 0$ or $r(0) \neq 0$), we perform a "small" perturbation.

In order to solve $L(h) = {}^{CF}_0 \mathcal{D}^\alpha_t h + \lambda_1 h' + \lambda_0 h$ and the IVP

$$
\begin{cases}
L(h)(t) = r(t) \\
h(0) \quad = h_0
\end{cases}
$$

with $h_0 \neq 0$, we consider $\tilde{r}(t) = r(t) - \lambda_0 h_0$ and, for small $\varepsilon > 0$, $\tilde{r}_\varepsilon(t)$ a $C^\infty(\mathbb{R})$ function on $(0, \varepsilon)$, that is null at the origin and coincides with $r(t)$ for $t > \varepsilon$ (see Fig. 3).

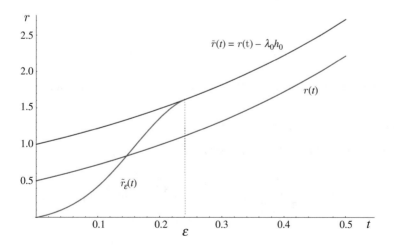

Fig. 3 Perturbation for the case $h(0) = h_0 \neq 0$

The solution h_ε to $L(h)(t) = \tilde{r}_\varepsilon(t)$ will be null at the origin and $h = h_\varepsilon + h_0$ satisfies the initial condition and

$$L(h_\varepsilon(t) + h_0) = {}^{CF}_0\mathcal{D}_t^\alpha h_\varepsilon(t) + \lambda_1 h_\varepsilon'(t) + \lambda_0 h_\varepsilon(t) + \lambda_0 h_0 = \tilde{r}_\varepsilon(t) + \lambda_0 h_0 \cong r(t)$$

Thus $h = h_\varepsilon + h_0$ is an approximate solution to the original IVP.

3.3   The Error

We comment on the error introduced in the different steps of the proposed numerical scheme.

First, we project and truncate the data r introducing two sources of error: we consider $r \cong \tilde{r} = \sum_{j=J_{min}}^{J_{max}} r_j$, $r_j \in W_j$, satisfying $r - \tilde{r} = e_r$ with $\|e_r\|_2 < \epsilon\|r\|_2 \simeq 0$, and, for each j, we perform a truncation by posing $r_j \cong \tilde{r}_j = \sum_{k \in \mathbb{K}_j} c_{jk}\psi_{jk}(t)$, $\mathbb{K}_j \subset \mathbb{Z}$, $|\mathbb{K}_j| = \eta_j < \infty$. The choice of J_{min}, J_{max} and $\mathbb{K}_j$ guarantees that the error introduced at this stage can be neglected. It can also be reduced by choosing a wider range for j and a larger $\mathbb{K}_j$.

Another source of error arises when considering $u_{jk} \in W_j$. It can be cut down posing the system simultaneously in more (all) levels.

Finally, the linear system (10) is posed and solved. The elements of $\mathcal{M}^j$ are the inner products $\langle u_{jk}, \psi_{jm} \rangle$. These integrals can be performed in the frequency domain taking advantage of the properties of the wavelets, i.e. on the compact subsets Ω_j:

$$\langle u_{jk}, \psi_{jm} \rangle = \frac{1}{2\pi} \int_{\Omega_j} \widehat{u}_{jk}(\omega)\widehat{\psi}_{jm}(\omega)d\omega$$

and can be computed with good precision.

We observe that, in order to calculate the wavelet coefficients b_k^j, it is not necessary to compute u_{jk}: we only need the values of $\langle u_{jk}, \psi_{jm} \rangle$.

Regarding the solution to (10), $\mathcal{M}^j$ can be arbitrarily approximated by band matrices:

Lemma $\mathcal{M}^j$ *is nearly a band matrix.*

Proof. From (8), there exist $N \in \mathbb{N}$ such that

$$H(\omega) = \frac{1}{2\pi} \left(\sum_{n=0}^{N} a_n \cos(\frac{n\omega}{2^j}) + i \sum_{n=1}^{N} b_n \sin(\frac{n\omega}{2^j}) \right) + \epsilon(\omega)$$

where $\epsilon(\omega)$ is an error that is small for large N. Then, from (9) ,

$$u_{jk}(t) \cong \frac{M(\alpha)}{2\pi(1-\alpha)} \int_{\Omega_j} \widehat{\psi}_{jk}(\omega) \left(\sum_{n=0}^{N} a_n \cos(\frac{n\omega}{2^j}) + i \sum_{n=1}^{N} b_n \sin(\frac{n\omega}{2^j}) \right) e^{i\omega t} d\omega.$$

We note that

$$a_n \cos(\frac{n\omega}{2^j}) \widehat{\psi}_{jk}(\omega) = \frac{a_n}{2}(e^{i\frac{n\omega}{2^j}} + e^{-i\frac{n\omega}{2^j}}) \widehat{\psi}_{jk}(\omega) = \frac{a_n}{2}(\widehat{\psi}_{j(k-n)}(\omega) + \widehat{\psi}_{j(k+n)}(\omega))$$

and

$$b_n \sin(\frac{n\omega}{2^j}) \widehat{\psi}_{jk}(\omega) = \frac{b_n}{2i}(e^{i\frac{n\omega}{2^j}} - e^{-i\frac{n\omega}{2^j}}) \widehat{\psi}_{jk}(\omega) = \frac{b_n}{2i}(\widehat{\psi}_{j(k-n)}(\omega) - \widehat{\psi}_{j(k+n)}(\omega)).$$

Consequently,

$$u_{jk}(t) \cong \frac{M(\alpha)}{1-\alpha} \left[a_0 \psi_{jk}(t) + \sum_{n=1}^{N} \left(\frac{a_n + b_n}{2} \psi_{j(k-n)}(t) + \frac{a_n - b_n}{2} \psi_{j(k+n)}(t) \right) \right]$$

and for $m \in \mathbb{K}_j, 0 \le m \le N,$ we can approximate the elements of the matrix by

$$\mathcal{M}_{km}^j = \langle u_{jk}, \psi_{jm} \rangle \cong \begin{cases} \frac{M(\alpha)}{1-\alpha} \frac{a_{k-m}+b_{k-m}}{2}, & \text{if } k-m > 0 \\ \frac{M(\alpha)}{1-\alpha} \frac{a_{k+m}-b_{k+m}}{2}, & \text{if } k-m < 0 \end{cases}$$

$$\mathcal{M}_{kk}^j = \langle u_{jk}, \psi_{jk} \rangle \cong \frac{M(\alpha)}{1-\alpha} a_0.$$

Since the inner products are zero for $m > N$, $\mathcal{M}^j$ is nearly a band matrix. ◇

In all the numerical experiments we performed, $\mathcal{M}^j$ was a diagonal dominant matrix, with good condition number, and the linear system was solved efficiently.

4   A Fractional Boundary Value Problem

As we mentioned in the Introduction, we can apply the proposed methodology to find approximate solutions to linear fractional boundary value problems (FBVP). We will explain the method on a specific problem: anomalous diffusion.

A fractional diffusion equation involving fractional temporal derivatives (see [33, 34]) is used to model, for example, dispersion phenomena in heterogeneous media. There are experimental results – like anomalous diffusion in porous, fractal or biological media, in turbulent plasma and in polymers, among others – that show that the mean square displacement of the particles must be considered to be proportional not to the time but to a power of time to fit the empirical data. This power may be less than unity (*subdiffusion*) or greater than 1 (*superdiffusion*), and the diffusion model can be expressed by a differential equation of the type

$$\mathcal{D}_t^\beta u(\mathbf{x}, t) - k\nabla^2 u(\mathbf{x}, t) = s(\mathbf{x}, t), \tag{11}$$

where β is a fractional order of derivation with respect to time ($0 < \beta < 1$ for subdiffusion and $\beta > 1$ for superdiffusion).

We will consider Caputo-Fabrizio temporal fractional derivatives in the diffusion equation (11). Computations for the case of CFD are similar.

For the 1D case (i.e. $\mathbf{x} \in \mathbb{R}$) we will solve the following boundary value problem of anomalous diffusion:

$$\begin{cases} {}^{CF}_0\mathcal{D}_t^\beta u(x,t) - \frac{\partial^2 u}{\partial x^2}(x,t) = s(x,t), & 0 < t < T, \quad 0 < x < L \\ u(x,0) = f(x) & 0 < x < L \\ \frac{\partial u}{\partial t}(x,0) = g(x) & 0 < x < L \\ u(0,t) = u(L,t) = 0 & t > 0 \end{cases} \tag{12}$$

where $\beta = 1 + \alpha$, $0 < \alpha < 1$ and ${}^{CF}_0\mathcal{D}_t^{1+\alpha}$ means ${}^{CF}_0\mathcal{D}_t^\alpha(u')$ (see [28]). The initial data f and g are supposed to be sufficiently smooth, with $g(0) = 0$. The source $s(x, t)$ is a smooth and causal function (i.e. $s(x, t) = 0$ for $t \le 0$).

Note that, for h regular enough, we have (see [28, 35, 36])

$$
{}^{CF}_0\mathcal{D}_t^{1+\alpha}h(t) = {}^{CF}_0\mathcal{D}_t^\alpha h'(t) = \frac{M(\alpha)}{1 - \alpha}\int_0^t h''(\tau)e^{-\frac{\alpha(t-\tau)}{1-\alpha}}\,d\tau.
$$

Integrating by parts we obtain

$$
{}^{CF}_0\mathcal{D}_t^{1+\alpha}h(t) = \frac{M(\alpha)}{1 - \alpha}\left[h'(t) - h'(0)e^{-\frac{\alpha t}{1-\alpha}} - \frac{\alpha}{1-\alpha}\int_0^t h'(\tau)e^{-\frac{\alpha(t-\tau)}{1-\alpha}}\,d\tau\right],
$$

that is

$$
{}^{CF}_0\mathcal{D}_t^{1+\alpha}h(t) = \frac{M(\alpha)}{1 - \alpha}[h'(t) - h'(0)e^{-\frac{\alpha t}{1-\alpha}}] - \frac{\alpha}{1-\alpha}{}^{CF}_0\mathcal{D}_t^\alpha h(t). \tag{13}
$$

Now we replace (13) in (12) and arrive to a time FPDE of order α, $0 < \alpha < 1$:

$$
\frac{M(\alpha)}{1 - \alpha}[u_t(x,t) - u_t(x,0)e^{-\frac{\alpha t}{1-\alpha}}] - \frac{\alpha}{1-\alpha}{}^{CF}_0\mathcal{D}_t^\alpha u(x,t) - u_{xx}(x,t) = s(x,t).
$$

Taking into account the initial conditions we have

$$
\frac{M(\alpha)}{1 - \alpha}u_t(x,t) - \frac{M(\alpha)}{1-\alpha}g(x)e^{-\frac{\alpha t}{1-\alpha}} - \frac{\alpha}{1-\alpha}{}^{CF}_0\mathcal{D}_t^\alpha u(x,t) - u_{xx}(x,t) = s(x,t)
$$

or

$$^{CF}_{\ 0}\mathcal{D}^{\alpha}_t u(x,t) + \frac{1-\alpha}{\alpha}u_{xx} - \frac{M(\alpha)}{\alpha}u_t = -\frac{1-\alpha}{\alpha}s(x,t) - \frac{M(\alpha)}{\alpha}g(x)e^{-\frac{\alpha t}{1-\alpha}}.$$

Let us define

$$\tilde{s}(x,t) = -\frac{1-\alpha}{\alpha}s(x,t) - \frac{M(\alpha)}{\alpha}g(x)e^{-\frac{\alpha t}{1-\alpha}}.$$

The resulting FBVP is

$$\begin{cases} ^{CF}_{\ 0}\mathcal{D}^{\alpha}_t u(x,t) + \frac{1-\alpha}{\alpha}u_{xx}(x,t) - \frac{M(\alpha)}{\alpha}u_t(x,t) = \tilde{s}(x,t), & 0 < t < T, \quad 0 < x < L \\ \quad\quad\quad\quad\quad\quad u(x,0) = f(x) & 0 < x < L \\ \quad\quad\quad\quad\quad\quad u(0,t) = u(L,t) = 0 & t > 0 \end{cases}.$$

$$(14)$$

We will construct an approximate solution to (14) as superposition of smooth functions (in the Schwartz class). As in the standard case (natural order PDE), we propose a solution to (14) by separating variables, and one of the resulting ODE will have fractional order.

If $u(x,t) = X(x)T(t)$ we pose the second order ODE

$$X'' - \nu X = 0$$

with boundary condition $X(0) = X(L) = 0$, and find $\nu = -(\frac{k\pi}{L})^2$, $X_k(x) = \sin(\frac{k\pi x}{L})$, for $k \in \mathbb{Z}$.

Now, for $u(x,t) = \sum_{k\geq 1} u_k(t)\sin(\frac{k\pi x}{L})$, and supposing that derivation and summation can be interchanged, we replace this last expression in (14) and obtain

$$\sum_{k\geq 1}[^{CF}_{\ 0}\mathcal{D}^{\alpha}_t u_k(t) - \frac{M(\alpha)}{\alpha}u'_k(t) - (\frac{k\pi}{L})^2\frac{1-\alpha}{\alpha}u_k(t)]\sin(\frac{k\pi}{L}x) = \tilde{s}(x,t). \quad (15)$$

Note that, if $u \in C^2(0,1) \times C^1(0,T)$, the derivatives $u_{xx}(x,t)$, $u_t(x,t)$ and $^{CF}_{\ 0}\mathcal{D}^{\alpha}_t u(x,t) = \frac{M(\alpha)}{1-\alpha}\int_0^t u_t(x,s)e^{-\frac{\alpha(t-s)}{1-\alpha}}ds$ are continuous functions in $(0,1) \times (0,T)$.

If $u^{**}_k(t)$, $u^{\#}_k(t)$ and $u^*_k(t)$ are, respectively, the Fourier coefficients of $u_{xx}(x,t)$, $u_t(x,t)$ and $^{CF}_{\ 0}\mathcal{D}^{\alpha}_t u(x,t)$ for each $t \in [0,T]$, it follows that $u^{**}_k(t) = \frac{-k^2\pi^2}{L^2}u_k(t)$, $u^{\#}_k(t) = u'_k(t)$ and $u^*_k(t) = {}^{CF}_{\ 0}\mathcal{D}^{\alpha}_t u_k(t)$. From (15) we have that the Fourier coefficients of $\tilde{s}$, $\tilde{s}_k(t) = 2\int_0^L \tilde{s}(r,t)\sin(\frac{k\pi}{L}r)dr$, must satisfy

$$^{CF}_{\ 0}\mathcal{D}^{\alpha}_t u_k(t) - \frac{M(\alpha)}{\alpha}u'_k(t) - (\frac{k\pi}{L})^2\frac{1-\alpha}{\alpha}u_k(t) = \tilde{s}_k(t)$$

Then, the functions u_k are solutions to the following IVP (similar to (7)):

$$\begin{cases} {}^{CF}_0\mathcal{D}^\alpha_t u_k(t) - \frac{M(\alpha)}{\alpha} u'_k(t) - (\frac{k\pi}{L})^2 \frac{1-\alpha}{\alpha} u_k(t) = \tilde{s}_k(t) \\ \qquad\qquad\qquad\qquad\qquad\qquad u_k(0) = f_k \end{cases} \qquad (16)$$

where $f_k = 2 \int_0^L f(\mu) \sin(\frac{k\pi}{L}\mu)d\mu$ are the Fourier coefficients of f.

Under the assumptions that $v(0) = 0$ and l causal, there is a unique solution in $C^1[0, T]$ for

$$L(v) = {}^{CF}_0\mathcal{D}^\alpha_t v + \lambda_0 v + \lambda_1 v' = l(t)$$

(see [32]) and we can approximate it by a smooth function: a linear combination of wavelets.

Explicit formula for the solution to (16) may also be found in some cases.

Note that, regarding the hypothesis on s, $\tilde{s}_k(0) = 2\int_0^L \tilde{s}(\mu, 0) \sin(\frac{k\pi}{L}\mu)d\mu$ might not be null because $\tilde{s}(\mu, 0) = -\frac{1-\alpha}{\alpha}s(\mu, 0) - \frac{M(\alpha)}{\alpha}g(\mu)$. In addition, from the initial conditions, we know that $u_k(0) = f_k$. If $f_k \neq 0$, we have to adapt the scheme in order to apply the same methodology, as explained previously.

Finally, $u(x, t) = \sum_{k \geq 1} u_k(t) \sin(\frac{k\pi x}{L})$.

5 Numerical Examples

In this section we show the performance of the proposed numerical approximation in some examples. The FPDE is

$$ {}^{CF}_0\mathcal{D}^{1+\alpha}_t u(x, t) - \frac{\partial^2 u}{\partial x^2}(x, t) = s(x, t), \qquad 0 < t < T, \quad 0 < x < L. $$

In Examples 1, 2 and 3 we build approximate solutions following the proposed technique for different initial and boundary conditions.

5.1 Example 1

Let us consider the following FBVP:

$$\begin{cases} {}^{CF}_0\mathcal{D}^{3/2}_t u(x, t) - u_{xx}(x, t) = v(t) \sin(-3\pi t) \sin(\frac{\pi}{10}x), 0 < x < 0, 0 < t < 16, \\ u(x, 0) = \sin(\frac{\pi}{5}x), \ \forall x \in [0, 10] \\ u_t(x, 0) = 0 \\ u(0, t) = u(10, t) = 0, \ \forall t \in [0, 16] \end{cases}$$

Table 1 Energy distribution of the data function $\tilde{s}_1$ and of the solution u_1

Level j	Energy of $\tilde{s}_1$ (%)	Energy of u_1 (%)	Frequency (ω)
1	0.0108	0.0927	[6.28, 12.5]
0	0.0068	0.0016	[3.14, 6.28]
−1	0.0167	0.0045	[1.57, 3.14]
−2	0.0504	0.0165	[0.78, 1.57]
−3	**0.3805**	**0.1270**	[0.39, 0.78]
−4	**0.5318**	**0.6321**	[0.19, 0.39]
−5	0.0057	0.1251	[0.09, 0.19]

where $v(t)$ is a smooth window in $[0, 16]$ and $(x, t) \in (0, 10) \times (0, 16)$. Following the steps described above, if $u(x, t) = \sum_{k \geq 1} u_k(t) \sin(\frac{k\pi x}{L})$, we only need to solve the (16) for $k = 1$ and arrive to the

$$\begin{cases} {}_{0}^{CF}\mathcal{D}_t^{1/2} u_1(t) - 2u_1'(t) - (\frac{\pi}{10})^2 u_1(t) = v(t)\sin(3\pi t) + (\frac{\pi}{10})^2, \, 0 < t < 16 \\ u_1(0) = 0 \end{cases}$$

In Table 1 the energy wavelet analysis by levels of the functions $\tilde{s}_1$ and of the solution u_1 is shown. The significant levels $j = -4, -3$ contain the 91% of $\tilde{s}_1$. For the reconstruction we consider levels $-1 \leq j \leq -5$. The resulting mean square error is $4.3866 * 10^{-6}$. We plot the exact u_1 vs. its approximation in Fig. 4 and the solution to the BVP in Fig. 5.

5.2   Example 2

We consider the same FPDE as in Example 1, but changing the initial condition by $u(x, 0) = \sin(\frac{\pi}{10}x)$, $\forall x \in [0, 10]$.

Separating variables we arrive to

$$\begin{cases} {}_{0}\mathcal{D}_t^{1/2} u_1(t) - 2u_1'(t) - (\frac{\pi}{10})^2 u_1(t) = v(t)\sin(3\pi t) = \tilde{s}_1(t), \, 0 < t < 16 \\ u_1(0) = 1 \end{cases}$$

As the initial condition on u_1 is not null, we perform the perturbation already described.

Table 2 contains the energy wavelet analysis by levels of the functions $\tilde{s}_1$ and of the solution u_1. The significant levels $j = -4, -3$ contain the 91% of $\tilde{s}_1$.

For the reconstruction we consider levels $-1 \leq j \leq -5$. The resulting mean square error is $1.8338 * 10^{-4}$.

In Figs. 6 and 7 we show the approximation for $u_1(t)$ versus the exact solution and the approximation for $u(x, t)$, respectively.

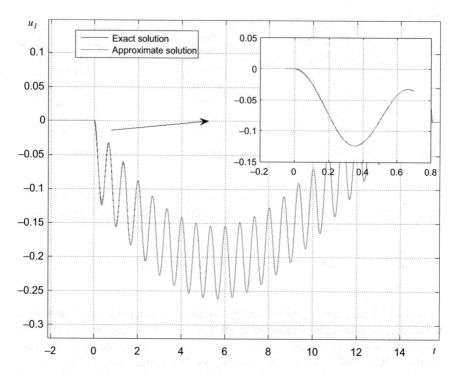

Fig. 4 $u_1(t)$ vs. $u_1(t)$ approx.

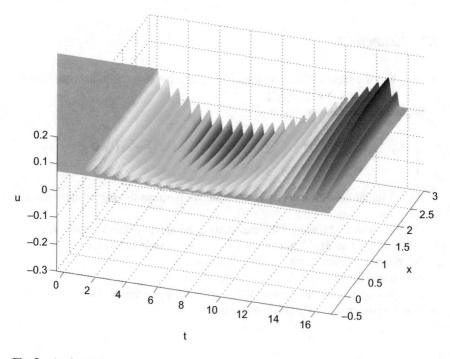

Fig. 5 $u(x, t)$ approx.

Table 2 Energy distribution of the data function $\tilde{s}_1$ and of the solution u_1

Level j	Energy of $\tilde{s}_1$ (%)	Energy of u_1 (%)	Frequency (ω)
1	0.0108	0.0210	[6.28, 12.5]
0	0.0068	0.0322	[3.14, 6.28]
−1	0.0167	0.0736	[1.57, 3.14]
−2	0.0504	0.1920	[0.78, 1.57]
−3	**0.3805**	**0.4747**	[0.39, 0.78]
−4	**0.5318**	**0.1978**	[0.19, 0.39]
−5	0.0057	0.0006	[0.09, 0.19]

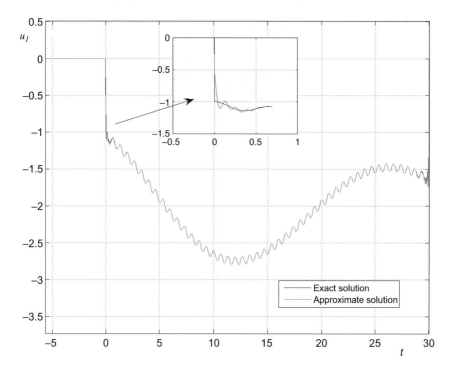

Fig. 6 $u_1(t)$ vs. $u_1(t)$ Approx.

5.3 Example 3

In order to evaluate the performance of the method, we obtain the solutions to (12) for different values of β approaching 2 and compare them with that of the wave equation, which corresponds to $\beta = 2$.

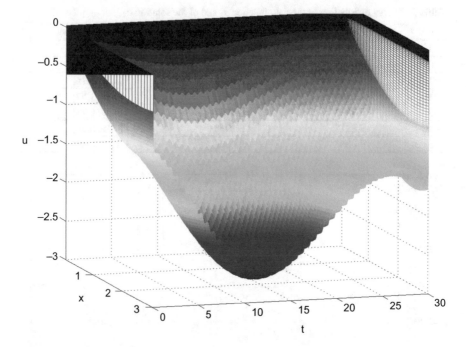

Fig. 7 Approx. $u(x, t)$ sol.

Consider

$$
\begin{cases}
{}^{CF}_{0}\mathcal{D}^{\beta}_{t}u(x, t) - u_{xx}(x, t) = v(t)\sin(-3\pi t)\sin(\tfrac{\pi}{10}x), \, 0 < x < 10, 0 < t < 32, \\
\quad\quad\quad u(x, 0) = 0, \, \forall x \in [0, 10] \\
\quad\quad\quad u_t(x, 0) = 0 \\
\quad u(0, t) = u(10, t) = 0, \, \forall t \in [0, 32]
\end{cases}
$$

where $v(t)$ is a smooth window in $[0, 32]$ and $(x, t) \in (0, 10) \times (0, 32)$, with $\beta = 1 + \alpha$ for different α. When $\alpha \to 1$ the solutions must resemble those of $\beta = 2$.

Following the steps described above, if $u(x, t) = \sum_{k \geq 1} u_k(t) \sin(\tfrac{k\pi x}{L})$, we only need to solve (16) for $k = 1$. We consider $\beta = 1.8, 1.9, 1.95$.

See Figs. 8 and 9 where we show, respectively, $u_1(t)$ and $u(x, t)$ for the different values of β.

On the other hand, if $\beta \to 1$ in (12), the behaviour of the solution u would tend to that of the classical diffusion equation.

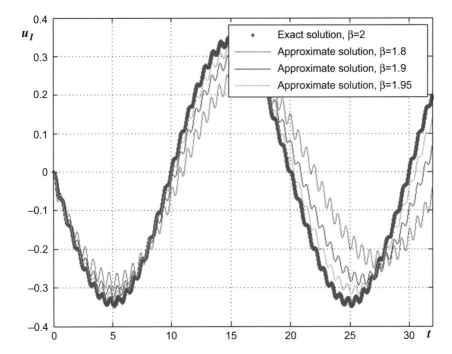

Fig. 8 $u_1(t)$ for different β

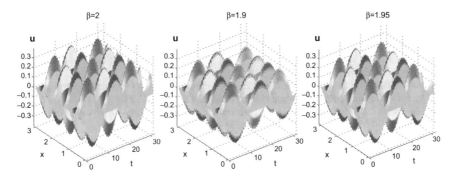

Fig. 9 $u(x, t)$ approx. for different β

6 Conclusions and Future Work

In this work we have adapted a numerical scheme to solve FODE, developed in previous works, to find approximate solutions to a FBVP. Using the proposed method, we built approximate solutions to an advection diffusion equation of order β, $1 < \beta < 2$. When β tends to 2, its behaviour looks like that of the solution to the standard wave equation. We have developed the calculations considering the CFFD but the CFD case is analogous. The same scheme can be proposed to solve different linear FPDE.

We intend to apply this methodology to solve inverse problems involving fractional models. Extensions to some nonlinear equations are also being studied.

Acknowledgments This work was partially supported by Universidad de Buenos Aires, under grant UBACyT20020170100350BA.

References

1. Kilbas, A., Srivastana, H.M., Trujillo, J.J.: Theory and Applications of Fractional Differential Equations. North Holland Mathematics Studies, vol. 204. Elsevier, Amsterdam (2006)
2. Miller, K., Ross, B.: An Introduction to the Fractional Calculus and Fractional Differential Equations. Wiley, New York (1993)
3. Oldham, K., Spanier, J.: The Fractional Calculus. Academic, New York/London (1974)
4. Podlubny, I.: Fractional Differential Equations. Academic, San Diego (1999)
5. Baleanu, D., Agarwal, R., Mohammadi, H., Rezapour, S: Some existence results for a nonlinear fractional differential equation on partially ordered Banach spaces. Bound. Value Probl. 112 (2013). https://doi.org/10.1186/1687-2770-2013-112
6. Baleanu, D., Diethelm, K., Scalas, E., Trujillo, J.: Fractional Calculus: Models and Numerical Methods. World Scientific Publishing, Singapore (2012)
7. Lin, S., Lu, C.: Laplace transform for solving some families of fractional differential equations and its applications. Adv. Differ. Equ. (2013). https://doi.org/10.1186/1687-1847-2013-137
8. Inc, M.: The approximate and exact solutions of the space- and time-fractional Burgers equations with initial conditions by variational iteration method. J. Math. Anal. Appl. **345**, 476–484 (2008)
9. Javed, I., Ahmadb, A., Hussaind, M., Iqbala, S.: Some Solutions of Fractional Order Partial Differential Equations Using Adomian Decomposition Method (2017). arXiv:1712.09207[math.NA]
10. Meerschaert, M., Tadjeran, C.: Finite difference approximations for two-sided space-fractional partial differential equations. Appl. Numer. Math. **56**(1), 80–90 (2006)
11. Momani, S., Odibat, Z.: Homotopy perturbation method for nonlinear partial differential equations of fractional order. Phys. Lett. A **365**, 345–350 (2007)
12. Yavuz, M., Ozdemir, N.: Comparing the new fractional derivative operators involving exponential and Mittag-Leffler kernel. Discret. Contin. Dyn. Syst. S **13**(3), 995–1006 (2020). https://doi.org/10.3934/dcdss.2020058
13. Mainardi, F.: Fractional calculus. In: Carpinteri, A., Mainardi, F. (eds.) Fractals and Fractional Calculus in Continuum Mechanics. International Centre for Mechanical Sciences. Courses and Lectures, vol. 378. Springer, Vienna (1997). https://doi.org/10.1007/978-3-7091-2664-6-7
14. Zhang, J., Zhang, X., Yang, B.: An approximation scheme for the time fractional convection diffusion equation. Appl. Math. Comput. **335**, 305–312 (2018). https://doi.org/10.1016/j.amc.2018.04.019
15. Mainardi, F., Paradisi, P.: Fractional diffusive waves. J. Comput. Acoust. **9**(4), 1417–1436 (2001). https://doi.org/10.1016/S0218-396X(01)00082-6
16. Povstenko, Y.: Fractional thermoelasticity problem for an infinite solid with a cylindrical hole under harmonic heat flux boundary condition. Acta Mech. **230**, 2137–2144. https://doi.org/10.1007/s00707-019-02401-2
17. Tenreiro Machado, J., Silva, M., Barbosa, R., Jesus, I., Reis, C., Marcos, M., Galhano, A.: Some applications of fractional calculus in engineering. Math. Probl. Eng. Article ID 639801, 34 (2010). https://doi.org/10.1155/2010/639801

18. Yu, Y., Perdikaris, P., Karniadakis, G.: Fractional modeling of viscoelasticity in 3D cerebral arteries and aneurysms. J. Comput. Phys. **323**, 219–242 (2016). https://doi.org/10.1016/j.jcp. 2016.06.038
19. Gómez-Aguilar, J., López-López, M., Alvarado-Martínez, V., Baleanu, D., Khan, H.: Chaos in a cancer model via fractional derivatives with exponential decay and Mittag-Leffler law. Entropy **19**(681), 19 (2017). https://doi.org/10.3390/e19120681
20. Ucar, S., Ucar, E., Ozdemir, N., Hammouch, Z.: Mathematical analysis and numerical simulation for a smoking model with Atangana-Baleanu derivative. Chaos Solitons Fractals **118**, 300–306 (2018). https://doi.org/10.1016/j.chaos.2018.12.003
21. Ozdemir, N., Ucar, E.: Investigating of an immune system-cancer mathematical model with Mittag-Leffler kernel. AIMS Math. **5**(2), 1519–1531 (2020). https://doi.org/10.3934/math. 2020104
22. Ozdemir, N., Ucar, S., Iskender, B.: Dynamical analysis of fractional order model for computer virus propagation with kill signals. Int. J. Nonlinear Sci. Numer. Simul. (2019). https://doi.org/ 10.1515/ijnsns-2019-0063
23. Yavuz, M., Ozdemir, N.: A different approach to the European option pricing model with new fractional operator. Math. Model. Nat. Phenom. **13**(1) (2018). https://doi.org/10.1051/mmnp/ 2018009
24. Yavuz, M., Ozdemir, N.: European vanilla option pricing model of fractional order without singular kernel. Fractal Fract. **2**(1), 3 (2018). https://doi.org/10.3390/fractalfract2010003
25. Fabio, M., Troparevsky, M.I.: Numerical solution to initial value problems for fractional differential equations. Progr. Fract. Differ. Appl. Int. J. **5**(3), 195–206 (2019). https://doi.org/ 10.18576/pfda/050302
26. Fabio, M., Troparevsky, M.I.: An inverse problem for the Caputo fractional derivative by means of the wavelet transform. Progr. Fract. Differ. Appl. Int. J. **4**(1), 15–26 (2018)
27. Caputo, M.: Linear models of dissipation whose Q is almost frequency independent, Part II. Geophys. J. R. Astr. Soc. **13**, 529–539 (1967)
28. Caputo, M., Fabrizio, M.: A new definition of fractional derivative without singular kernel. Progr. Fract. Differ. Appl. **1**(2), 73–85 (2015)
29. Fabio, M., Serrano, E.: Infinitely oscillating wavelets and an efficient implementation algorithm based on the FFT. Revista de Matemática: Teoría y Aplicaciones **22**(1), 61–69 (2015). CIMPA – UCR ISSN: 1409-2433 (PRINT), 2215-3373 (ONLINE)
30. Mallat, S.: A Wavelet Tour of Signal Processing. Academic/Elsevier, Boston/Amsterdam (2009)
31. Meyer, Y.: Ondelettes et Operateurs II: Operatteurs de Calderon Zygmund. Hermann et Cie, Paris (1990)
32. Troparevsky, M.I., Fabio, M.: Approximate solutions to initial value problems with combined derivatives [in Spanish]. Mecánica Computacional **XXXVI**(11), 449–459 (2018)
33. Liu, F., Zhuang, P., Burrage, K.: Numerical methods and analysis for a class of fractional advection-dispersion models. Comput. Math. Appl. **64**, 2990–3007 (2012)
34. Xu, Y., He, Z., Xu, Q.: Numerical solutions of fractional advection-diffusion equations with a kind of new generalized fractional derivative. Int. J. Comput. Math. (2013). https://doi.org/10. 1080/00207160.2013.799277
35. Al Salti, N., Karimov, E., Kerbal, S.: Boundary value problems for fractional heat equation involving Caputo-Fabrizio derivative. NTMSCI **4**, 79–89 (2016)
36. Losada, J., Nieto, J.: Properties of a new fractional derivative without singular kernel. Prog. Fract. Differ. Appl. **1**(2), 87–92 (2015)

Wavelet B-Splines Bases on the Interval for Solving Boundary Value Problems

Lucila Calderón, María T. Martín, and Victoria Vampa

Abstract The use of multiresolution techniques and wavelets has become increasingly popular in the development of numerical schemes for the solution of differential equations. Wavelet's properties make them useful for developing hierarchical solutions to many engineering problems. They are well localized, oscillatory functions which provide a basis of the space of functions on the real line. We show the construction of derivative-orthogonal B-spline wavelets on the interval which have simple structure and provide sparse and well-conditioned matrices when they are used for solving differential equations with the wavelet-Galerkin method.

1 Introduction

In recent years, wavelet methods have been developed as a new powerful tool for the numerical solution of some boundary value problems.

Wavelets and multiresolution analysis (MRA) provide a robust and accurate alternative to traditional methods for solving differential equations. Their advantage is appreciated when they are applied to problems having localized singular behavior. The solution is approximated by an expansion of scaling functions and wavelets, with the convenience that multiscale and localization properties can be exploited. The choice of the wavelet basis is governed by several factors including the desired order of numerical accuracy and computational effort.

In some cases multiscale bases are combined with finite element methods, and adaptive refinement strategies are designed (Chen et al. [1] and Bindal et al. [2]). Other authors applied adaptive procedures in wavelet collocation methods, as the method introduced by Cai and Kumar et al. [3, 4]. Wavelet-Galerkin methods using variational equations is a good alternative, producing an efficient regularization

L. Calderón · M. T. Martín · V. Vampa (✉)
Facultad de Ingeniería, Departamento de Ciencias Básicas, Universidad Nacional de La Plata,
La Plata, Argentina
e-mail: lucila.calderon@ing.unlp.edu.ar; victoria.vampa@ing.unlp.edu.ar

© The Author(s), under exclusive license to Springer Nature Switzerland AG 2021
J. P. Muszkats et al. (eds.), *Applications of Wavelet Multiresolution Analysis*,
SEMA SIMAI Springer Series 4, https://doi.org/10.1007/978-3-030-61713-4_2

action: in weak formulations for a given equation, the approximating functions can be relatively less regular and easier to construct [5].

To obtain high precision results, it is important that the associated system matrix, known as the *stiffness matrix*, be a sparse matrix with a small condition number. So, the choice of a wavelet basis satisfying some mathematical requirements is of great importance for the good performance of the method.

Wavelets on the real line are not suitable in applications which are defined on bounded domains, as the problem of solving differential equations numerically. Therefore it is necessary to adapt them. Wavelet bases on a bounded interval are usually constructed from wavelets on the real line. The main idea is to retain most of the inner functions, i.e., the scaling functions and wavelets whose support is contained in the interval, and to construct appropriate boundary scaling functions and wavelets separately. Properties such as smoothness, local support, and polynomial exactness of basis functions should be preserved.

Many constructions of cubic spline wavelets or multiwavelet bases on the interval have been proposed in recent years. Jia et al. [6] designed biorthogonal multiwavelets adapted to the interval [0, 1] based on Hermite cubic splines. They developed a pair of spline wavelets to solve the Sturm-Liouville equation with Dirichlet boundary conditions adapted to the interval [0, 1]. The wavelets at different levels are orthogonal with respect to the inner product $\langle u', v' \rangle$ rather than $\langle u, v \rangle$. The stiffness matrix is sparse, and its condition number is uniformly bounded.

On the other hand, Vampa et al. [7] have applied a spline-cubic-wavelet basis adapted to the interval with good results. In their work a modified wavelet-Galerkin method using B-spline scaling functions to solve boundary value problems is presented. This proposal combines variational equations with a collocation scheme and gives an approximation at an initial scale. Later, in [8] a refinement process using wavelets is developed, and the approximation is improved recursively with minimal computational effort. A disadvantage of this construction is the large condition number of the stiffness matrices.

Later, Cerna et al. [9] proposed several constructions of cubic spline-wavelet bases. They presented different constructions of stable cubic spline-wavelet bases on the interval. Quantitative properties of constructed cubic spline-wavelet and multiwavelet bases are studied.

Due to their desirable properties, such as sparse stiffness matrices and small condition numbers, constructions of wavelet bases, whose *mth*-order derivatives are orthogonal among different levels, are of particular interest and importance in computational mathematics. In a general context, a theoretical study over this construction can be found in [10].

In the present work, we propose the construction of a cubic spline-wavelet basis with compact support and first derivatives functions orthogonal between the different scales. This inner product leads to a sparse stiffness matrix with a condition number uniformly bounded. This is a very important advantage of the proposed method.

The structure of the paper is as follows: in Sect. 2 we introduced a brief description of a wavelet-Galerkin method to solve a second-order linear differential operator. The review of the concept of wavelet bases, multiresolution analysis (MRA) structure on the interval, basic properties of B-splines functions and cubic B-splines subspaces are presented in Sect. 3. Section 4 contains the technical details of a construction of wavelet B-splines bases. In Sect. 5 they are applied as testing for efficient solution of a differential equation. Finally, some concluding remarks are made in Sect. 6.

2   Wavelet-Galerkin Method

We consider the following one-dimensional linear boundary value problem on the interval $I = [0, 1]$:

$$Lu(x) = -\frac{d}{dx}\left(p(x)\frac{du}{dx}\right) + q(x)u(x) = f(x) \tag{1}$$

$$u(0) = u(1) = 0,$$

where $p(x)$, $q(x)$, and $f(x)$ are continuous functions on I and u is a function in certain Hilbert space V. If Eq. (1) cannot be solved exactly, one has to rely on approximation methods. We seek an approximation $\tilde{u}$ of u which lies in a certain finite-dimensional subspace $\tilde{V} \subset V$.

Let $\langle \cdot, \cdot \rangle$ be the inner product of the space V. Note that $a(u, v) = \langle Lu, v \rangle$ defines a bilinear form on $V \times V$, so that the variational or weak formulation corresponding to the problem Eq. (1) is to seek $u \in V$, such that

$$a(u, v) = \langle f, v \rangle \quad \forall v \in V. \tag{2}$$

The analogous finite-dimensional problem is to find $\tilde{u} \in \tilde{V}$ such that

$$a(\tilde{u}, \tilde{v}) = \langle f, \tilde{v} \rangle \quad \forall \tilde{v} \in \tilde{V}. \tag{3}$$

It is well known that if $a(\cdot, \cdot)$ is continuous, V-elliptic and $\langle f, v \rangle$ is a continuous linear form in V, both problems Eqs. (2) and (3) have a unique solution (Lax-Milgram theorem [11]). From Céa's lemma [11] the following error bounds are valid:

$$\|u - \tilde{u}\|_V^2 \leq \frac{C}{\gamma} inf_{v \in \tilde{V}} \|u - v\|_V^2, \tag{4}$$

where C and γ are constants corresponding to continuity and coercivity of the bilinear form $a(.,.)$, and if h is a measure of the partition of I considered, then

$$\|u - \tilde{u}\|_V^2 \le Ch^r |u|_{H^{r+1}}^2, \tag{5}$$

where r depends on the regularity of the solution.

Going back to Eq. (1), integrating by parts $\langle Lu, v \rangle$, the associated bilinear form is

$$a(u, v) = \int_0^1 (p(x)u'(x)v'(x) + q(x)u(x)v(x)) \, dx, \tag{6}$$

for u and $v \in V^0 \subset L_2(I)$, the subspace of functions with homogeneous boundary conditions. Let $\{\Phi_1, \Phi_2, \ldots, \Phi_N\}$ a basis of $\tilde{V}$ and the approximate solution of the given equation be $\tilde{u} = \sum_{k=1}^N \alpha_k \Phi_k$. Replacing in Eq. (3) we have to determine α_k in a way that $\tilde{u}$ behaves as if it is a true solution in $\tilde{V}$, i.e.,

$$\sum_{k=1}^N \alpha_k \, a(\Phi_k, \Phi_n) = \langle f, \Phi_n \rangle, \quad n = 1, 2, \ldots, N. \tag{7}$$

We then arrive at the problem of solving a matrix equation

$$A\alpha = b, \tag{8}$$

where $A(n, k) = a(\Phi_k, \Phi_n)$, $b_n = \langle f, \Phi_n \rangle$, and $\alpha = (\alpha_k)$.

Condition Number of a Matrix

It is known that a linear system $AX = Y$ has a unique solution X for every Y if the square matrix A is invertible. It is often observed that for two close values of Y and Y', X and X' are far apart. Such a linear system is called *badly conditioned*. Thus data Y is expected to be fairly accurate. *Condition number* of A is given by

$$cond(A) = \| A \| \| A^{-1} \|, \quad cond(A) \ge 1, \tag{9}$$

($\| \cdot \|$ is a matrix norm) and when A is symmetric, in norm 2 is

$$cond(A) = \frac{\max_i | \lambda_i(A) |}{\min_i | \lambda_i(A) |}, \tag{10}$$

where $\lambda_i(A)$ are matrix A eigenvalues. $cond(A)$ is a measure of the stability of the linear system under perturbation of the data Y.

For computational aspects, it is convenient to have a sparse matrix A, i.e., with a high proportion of entries 0, with a low condition number, and basis functions with a small support, regularity, and orthogonality. It is also desirable that the basis functions should be simple to evaluate, differentiate, and integrate. Finally, one wants the scheme to be refinable to allow that the approximation $\tilde{u}$ can be improved, modifying recursively the subspace $\tilde{V}$. If the basis functions Φ_k are generated from dilations and translations of a mother generating function, calculations become simpler. This suggests considering a MRA structure. Furthermore, if self-similarity given by scale relations is satisfied, a hierarchical approximation to the exact solution is obtained, and it is possible to refine and improve the accuracy of the approximate solution.

3   Wavelet Analysis on the Interval

MRA schemes [12] provide a powerful mathematical tool for function approximation and multiscale representation of the solution of differential equations corresponding to the problem in Eq. (1). It is important to point out that, as these structures are generally defined on the whole real line, they must be adequately restricted to the interval I where the differential problem is formulated.

Many constructions of wavelet bases on the interval have been proposed. In [13] a family of orthonormal wavelets on a bounded interval by restricting Daubechies scaling functions and wavelets to [0, 1] was constructed by Meyer. Later, Chui and Quak [14] obtained spline-wavelet bases of $L_2[0, 1]$.

When a MRA on an interval is proposed, the usual strategy is to start from a MRA on $L_2(\mathbb{R})$ and then use a finite set of suitable translates $\varphi_{j,k}$ of the original scaling function and a finite set of specially constructed boundary scaling functions.

3.1   Multiresolution Analysis

As described by Chui [12], a MRA on $L_2(\mathbb{R})$ consists of a sequence of embedded closed subspaces $V_j \subset L_2(\mathbb{R})$, $j \in \mathbb{Z}$,

$$\cdots \subset V_{-2} \subset V_{-1} \subset V_0 \subset V_1 \subset V_2 \subset \cdots$$

that satisfies several properties and typically is constructed by first identifying the subspace V_0 and the scaling function ϕ. Denoting by

$$\phi_{j,k}(x) := 2^{j/2}\phi(2^j x - k) , \tag{11}$$

for each $j \in \mathbb{Z}$, the family $\{\phi_{j,k} : k \in \mathbb{Z}\}$ is a basis of V_j. Associated with the scaling function ϕ, there exists a function ψ called the *mother wavelet* such that the collection $\{\psi(x-k), k \in \mathbb{Z}\}$ is a Riesz basis [12] of W_0, the orthogonal complement of V_0 in V_1. If we consider

$$\psi_{j,k}(x) := 2^{j/2}\psi(2^j x - k) \,, \tag{12}$$

for each $j \in \mathbb{Z}$, the family $\{\psi_{j,k} : k \in \mathbb{Z}\}$ is a basis of W_j, the orthogonal complement of V_j in V_{j+1}. It is noteworthy that wavelets allow the refinement of the representation space taking into account that

$$V_{j+1} = V_j \oplus W_j. \tag{13}$$

3.2 MRA on the Interval

As it was mentioned above, multiresolution structures in $L_2(\mathbb{R})$ have to be restricted to $L_2(I)$, to solve boundary value problems on I (see [15] and [16]). If Haar bases are considered for $L_2(\mathbb{R})$, it suffices to take the restrictions of these functions to I. Things are not so trivial when one starts from smoother wavelets on the line. It is not clear a priori how to adapt the functions in such a way that an orthogonal basis is obtained. Several solutions have been proposed for this problem. A first solution is to extend the functions supported on I to the whole line by making them vanish for $x \notin I$. This approach may introduce a discontinuity at the edges, and consequently, large wavelet coefficients are obtained near the edges and too many wavelets are used. Another alternative consists in periodizing, but, unless the function itself is already periodic, it again introduces a discontinuity.

Consequently, restriction to I entails some changes in the concepts of a MRA. The aim is to produce Riesz bases for the spaces V_j consisting of a finite family of translates of the original scaling function $\phi_{j,k}$ and a finite family of special boundary scaling functions and to produce the bases of the complementary subspaces W_j consisting of a finite set of translates of the wavelet function $\psi_{j,k}$ and a finite set of special boundary wavelets.

In this work, a MRA on the interval with B-splines as scaling functions is described, and it is constructed using orthogonality conditions in a way similar as a MRA in $L_2(\mathbb{R})$.

3.2.1 B-Spline Subspaces

Spline wavelets are extremely regular and usually symmetric or antisymmetric. They can be designed to have compact support, and they have explicit expressions which facilitate not only theoretical formulation but also numerical implementations with a computer, see [15] and [17].

Let us consider B-spline functions of order $m + 1$, that is, connected piecewise polynomials of degree m having $m - 1$ continuous derivatives. The joining points are called knots, and they are typically equally spaced and positioned at the integers.

These functions can be defined recursively by convolutions [12]:

$$\varphi_1(x) = \chi_{[0,1]}(x),$$

$$\varphi_{m+1}(x) = \varphi_m * \varphi_1(x) \tag{14}$$

and constitute the scaling functions of the MRA structure.

Among many properties that B-splines have, the most important ones for our method are the following:

- Two-scale relation

$$\varphi_{m+1}(x) = 2^{-m} \sum_{k=0}^{m+1} \binom{m+1}{k} \varphi_{m+1}(2x - k). \tag{15}$$

- Differentiation

$$\frac{d^k}{dx^k} \varphi_{m+1}(x) = \Delta^k \varphi_{m+1-k}(x), \tag{16}$$

where Δ^k is the k-order difference operator and $1 \leq k \leq m - 1$, i.e., corresponds to a reduction of the spline degree by k.

- Inner products

$$\int_{\mathbb{R}} \varphi_{m+1}(x - k) \, \varphi_{n+1}(x - l) \, dx = \varphi_{m+n+2}(n + 1 + l - k), \tag{17}$$

i.e., simple evaluations of higher-order splines at integer points.

This property is obtained from the convolution product and is useful in weak formulations of differential problems.

In the B-spline MRA, V_0 is the subspace generated by the translates of the *scaling function* φ_{m+1} and for each $j \in \mathbb{Z}$, the family $\{\varphi_{m+1,j,k} : k \in \mathbb{Z}\}$ where

$$\varphi_{m+1,j,k}(x) := 2^{j/2} \varphi_{m+1}(2^j x - k) , \tag{18}$$

is a basis of V_j [15, 16]. These subspaces V_j constitute a MRA on $L_2(\mathbb{R})$.

3.2.2 Scaling Cubic B-Spline Subspaces

In this section, we introduce a cubic B-spline basis on the interval satisfying Dirichlet boundary conditions. This construction is based on the spline-wavelet bases defined by Chui and Quak in [14]. The adaptation of these bases to boundary conditions can be found in [19].

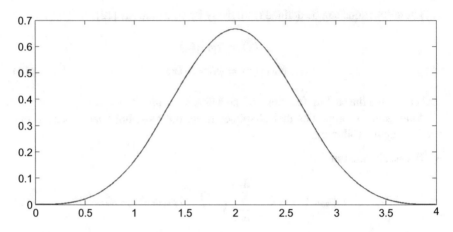

Fig. 1 Scaling function φ_4

In this work, we use B-splines of order $m = 3$. As they are C^2 functions, a hierarchical approximation of the solution for the second-order problem Eq. (1) can be obtained, and accurate results can most likely be expected [18].

In the cubic B-spline MRA framework, the scaling function φ_4 has support on $[0, 4]$ (Fig. 1), and $\{\varphi_{4,j,k}(x) := 2^{j/2}\varphi_4(2^j x - k) : k \in \mathbb{Z}\}$ is a basis of V_j.

It can be written explicitly as

$$\varphi_{3+1}(x) = \begin{cases} \dfrac{x^3}{6}, & x \in [0, 1] \\ -\dfrac{x^3}{2} + 2x^2 - 2x + \dfrac{2}{3}, & x \in [1, 2] \\ \dfrac{x^3}{2} - 4x^2 - 10x - \dfrac{22}{3}, & x \in [2, 3] \\ \dfrac{(4 - x)^3}{6}, & x \in [3, 4] \end{cases}. \tag{19}$$

To simplify the notation we call $\varphi(x) = \varphi_4(x)$.

Consider two boundary functions presented by Čërná et al. in the article [9]: φ_{b_1} y φ_{b_2}. They are piecewise cubic polynomials:

$$\varphi_{b_1}(x) = \begin{cases} \dfrac{7x^3}{4} - \dfrac{9x^2}{2} + 3x, & x \in [0, 1] \\ \dfrac{(2 - x)^3}{4}, & x \in [1, 2] \end{cases} \tag{20}$$

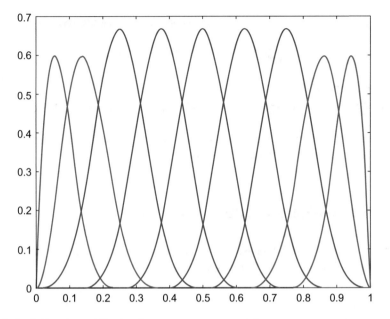

Fig. 2 Basis Functions of V_j, $j = 3$

and

$$\varphi_{b_2}(x) = \begin{cases} -\dfrac{11x^3}{12} + \dfrac{3x^2}{2}, & x \in [0, 1] \\ \dfrac{7x^3}{12} - 3x^2 + \dfrac{9x}{2} - \dfrac{3}{2}, & x \in [1, 2] \\ \dfrac{(3-x)^3}{4}, & x \in [2, 3] \end{cases}$$
(21)

If $\varphi_{j,k}(x) := 2^{j/2}\varphi(2^j x - k)$, for $j \in \mathbb{Z}$, the families

$$\Phi_j^{inn} = \left\{ \varphi_{j,k}(x) : k = 0, 1, \dots, 2^j - 4 \right\},$$
(22)

correspond to *inner scaling functions* and

$$\Phi_j^{bound} = \left\{ \varphi_{b_1}(2^j x), \varphi_{b_2}(2^j x), \varphi_{b_2}(2^j(1-x)), \varphi_{b_1}(2^j(1-x)) \right\},$$
(23)

are *boundary scaling functions*.

In Fig. 2 you can see inner and boundary scaling functions.

Now, considering the families Eqs. (22) and (23), the scaling space V_j is

$$V_j = \text{span } \Phi_j, \quad \text{where} \quad \Phi_j = \Phi_j^{inn} \cup \Phi_j^{bound}. \tag{24}$$

($\varphi_{j,k}$ are normalized so that $\| \varphi'_{j,k} \|_{L^2[0,1]} = 1$).

The dimension of the spaces V_j is $2^j + 1$, and they constitute a MRA on $L_2[0, 1]$ [19].

In the next section, the construction of Wavelet spaces W_j taking into account the decomposition Eq. (13) will be described.

4 Wavelet B-Splines: Orthogonal Basis

In the following, we build a basis for the wavelet spaces W_j with an orthogonality requirement, proposing a *mother wavelet* $\psi \in W_0$.

4.1 Construction of a Mother Wavelet

As $W_0 \subset V_1$, there exists a $\{d_k\}$ sequence such that

$$\psi(x) = \sum_{k \in \mathbb{Z}} d_k \, \varphi(2x - k), \quad x \in \mathbb{R}. \tag{25}$$

The coefficients $\{d(k)\}$ must be found such that the *orthogonality requirement*

$$\langle \psi'(x), \varphi'(x - l) \rangle = 0 \quad \forall l \in \mathbb{Z}, \tag{26}$$

is satisfied.

Fixed $l \in \mathbb{Z}$, this means

$$\langle \psi'(x), \varphi'(x - l) \rangle = 2 \left\langle \sum_{k \in \mathbb{Z}} \left[d_k \varphi'(2x - k) \right], \varphi'(x - l) \right\rangle$$

$$= 2 \sum_{k \in \mathbb{Z}} d_k \langle \varphi'(2x - k), \varphi'(x - l) \rangle. \tag{27}$$

Considering the intersection of the supports of scaling functions, index k takes only values $2l - 4 < k < 2l + 8$.

So we obtain

$$\langle \psi'(x), \varphi'(x-l) \rangle = 2 \sum_{k=2l-3}^{2l+7} d_k \langle \varphi'(2x-k), \varphi'(x-l) \rangle. \tag{28}$$

Rewriting the two-scale relation Eq. (15) as $\varphi(x) = \sum_{n=0}^{4} h_n \varphi(2x-n)$ and using properties of B-splines, the terms in the sum of Eq. (28) have the following expression:

$$\langle \varphi'(2x-k), \varphi'(x-l) \rangle = -2 \sum_{n=0}^{4} h_n \varphi_8''(4+2l+n-k). \tag{29}$$

Hence,

$$\langle \psi'(x), \varphi'(x-l) \rangle = -4 \sum_{k=2l-3}^{2l+7} d_k \sum_{n=0}^{4} h_n \varphi_8''(4+2l+n-k). \tag{30}$$

It remains to find d_k values. If we call

$$q_1(z) := \sum_{l \in \mathbb{Z}} d_{2l+1} z^{2l+1}, \qquad q_2(z) := \sum_{l \in \mathbb{Z}} d_{2l} z^{2l},$$

the orthogonality condition Eq. (26) is

$$B(z) (q_1(z) \; q_2(z))^T = 0.$$

where

$$(B(z))^T = \begin{bmatrix} -\dfrac{1}{240} z^7 - \dfrac{39}{80} z^5 + \dfrac{59}{120} z^3 + \dfrac{59}{120} z - \dfrac{39}{80} z^{-1} - \dfrac{1}{240} z^{-3} \\[2mm] -\dfrac{7}{60} z^6 - \dfrac{8}{15} z^4 + \dfrac{13}{10} z^2 - \dfrac{8}{15} - \dfrac{7}{60} z^{-2} \end{bmatrix}.$$

One solution is:

$$\begin{bmatrix} q_1(z) \\ q_2(z) \end{bmatrix} = \begin{bmatrix} -28\,z^5 - 184\,z^3 - 28\,z^1 \\ z^6 + 119\,z^4 + 119\,z^2 + 1 \end{bmatrix},$$

and therefore, the wavelet ψ is given by

$$\psi(x) = \sum_{k=0}^{6} d_k \varphi(2x-k), \quad x \in \mathbb{R}, \tag{31}$$

with $[d_0, d_1, \ldots, d_6] = [1, -28, 119, -184, 119, -28, 1]$.

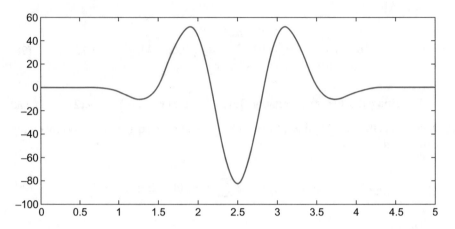

Fig. 3 Wavelet

$\psi(x)$ is supported on $[0, 5]$, it satisfies the orthogonality above conditions. Moreover $\psi(x)$ is symmetric (Fig. 3).

4.2 Wavelet Basis

We propose a suitable basis for the W_j spaces, considering two *boundary wavelets* ψ_{b_1}, $\psi_{b_2} \in W_0$ that are defined by Cěrná et al. [9]:

$$\psi_{b_1}(x) = c_0^{b_1} \varphi_{b1}(2x) + c_1^{b_1} \varphi_{b_2}(2x) + \sum_{k=2}^{4} c_k^{b_1} \varphi(2x - k + 2), \qquad (32)$$

$$\psi_{b_2}(x) = c_0^{b_2} \varphi_{b_1}(2x) + c_1^{b_2} \varphi_{b_2}(2x) + \sum_{k=2}^{6} c_k^{b_2} \varphi(2x - k + 2), \qquad (33)$$

where

$$\left[c_0^{b_1}, c_1^{b_1}, \ldots, c_4^{b_1} \right] = \left[\frac{939}{70}, \frac{-393}{20}, \frac{6233}{560}, -4, 1 \right],$$

$$\left[c_0^{b_2}, c_1^{b_2}, \ldots, c_6^{b_2} \right] = \left[\frac{1444}{953}, \frac{1048}{1871}, \frac{-1340}{209}, \frac{545}{48}, \frac{-6839}{655}, 7, -3 \right].$$

Boundary wavelets ψ_{b_1}, ψ_{b_2} have supports on $[0, 3]$ and $[0, 4]$, respectively. They have two vanishing moments and satisfy the orthogonality condition Eq. (26) (Fig. 4).

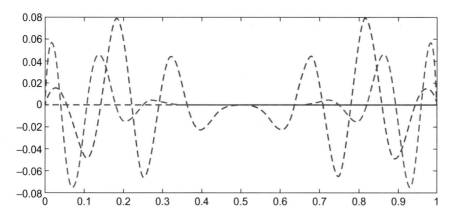

Fig. 4 Boundary wavelets of W_j, $j = 3$

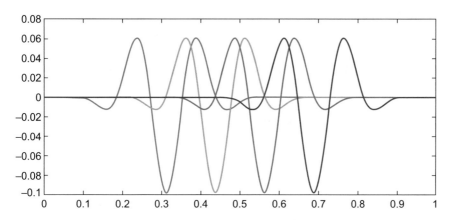

Fig. 5 Inner wavelets of W_j, $j = 3$

Using those functions, the set of *boundary wavelets* (Fig. 4) is defined:

$$\Psi_j^{bound} = \left\{ \psi_{b_1}(2^j x), \psi_{b_2}(2^j x), \psi_{b_2}(2^j(1-x)), \psi_{b_1}(2^j(1-x)) \right\}. \tag{34}$$

Note that as $V_{j+1} = V_j \oplus W_j$, the dimension of W_j is 2^j. Thus, a basis for these spaces is

$$\Psi_j = \Psi_j^{inn} \cup \Psi_j^{bound}, \tag{35}$$

where Ψ_j^{inn} is the set of *inner wavelets* (Fig. 5),

$$\Psi_j^{inn} = \left\{ \psi_{j,k} : k = 0, 1, \ldots, 2^j - 5 \right\}, \tag{36}$$

and $\psi_{j,k}(x) := 2^{j/2}\psi(2^j x - k)$, for each $j \in \mathbb{Z}$.

The functions in Ψ_j are normalized so that $\| \psi'_{j,k} \|_{L_2(0,1)} = 1$.

Remark 1 Due to $V_j \cap W_j = \{0\}$ and Eq. (13),

$$\dim(V_j + W_j) = \dim V_j + \dim W_j = 2^{j+1} + 1 = \dim(V_{j+1}). \qquad (37)$$

Thus,

$$V_{j+1} = V_{j_0} \oplus W_{j_0} \oplus W_{j_0+1} \ldots \oplus W_j, \quad \text{for } j_0 \geq 3.$$

For $J > j_0$, a wavelet basis for V_{J+1} is,

$$G_J = \Phi_{j_0} \cup \bigcup_{j=j_0}^{J} \Psi_j = \{g_1, g_2, \ldots, g_{2^{J+1}+1}\}, \qquad (38)$$

where $g_i \in \Phi_{j_0}$ for $i = 1, 2, \ldots, 2^{j_0} + 1$ and $g_i \in \Psi_j$ for $i = 2^{j_0} + 2, \ldots, 2^{J+1} + 1$ and $j = j_0 \ldots, J$.

Remark 2 If $v \in V_{j_0}$, $w_j \in W_j$, from the orthogonality condition Eq. (26) it is true that

$$\langle v', w'_{j_1} \rangle = 0,$$

$$\langle w'_{j_1}, w'_{j_2} \rangle = 0 \quad \text{for } j_1 \neq j_2. \qquad (39)$$

5   Numerical Example

Consider the following problem:

$$\begin{cases} -u'' = f & \text{on } (0, 1) \\ u(0) = u(1) = 0 \end{cases}, \qquad (40)$$

with $f(x) = (70\pi)^2 \sin(70\pi x) - \pi^2 \cos\left(\pi x + \dfrac{\pi}{2}\right)$.

Substitution of the approximation,

$$u_{J+1} = \sum_{i=1}^{2^{J+1}+1} \alpha_i \, g_i,$$

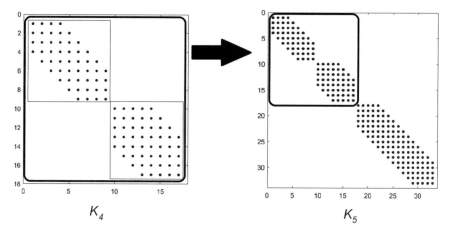

Fig. 6 Structure of matrices K_J, $J = 4, 5$

using the basis Eq. (38) into the weak formulation, Eq. (3) results in

$$\sum_{i=1}^{2^{J+1}+1} \alpha_i \left(\int_0^1 g_i'(x) g_l'(x) \, dx \right) = \int_0^1 f(x) g_l(x) \, dx \quad \forall l \in \{1, 2, \ldots, 2^{J+1} + 1\}.$$

or, in matrix form

$$K_J \, \alpha = R, \tag{41}$$

where K_J is the stiffness matrix,

$$\mathbf{K}_J := \left\langle g_i', g_j' \right\rangle_{1 \le l, i \le 2^{J+1}+1}. \tag{42}$$

This system of linear algebraic equations is solved for α, the vector of $2^{J+1} + 1 \times 1$ parameters.

As a consequence of the orthogonality requirement, the matrix K_J is sparse and each block is diagonal (Fig. 6). The condition number $cond(K_J) = \dfrac{\lambda_{max}}{\lambda_{min}}$ with respect to 2-norm is uniformly bounded. This assertion is confirmed by numerical computation of the condition number of the matrix K_J for $J = 3, \ldots, 9$ (see Table 1).

The exact solution of the problem is

$$u(x) = \sin(70 \pi x) - \cos\left(\pi x + \frac{\pi}{2}\right). \tag{43}$$

For $J = 1, 2, \ldots$, let $e_J := \dfrac{\|u_{J+1} - u\|}{\|u\|}$ the approximation relative errors.

Although the exact solution is very oscillatory, good convergence results were obtained, which are shown in Table 2 and Fig. 7.

Table 1 Condition number of K_J

J	3	4	5	6	7	8	9
λ_{max}	1.6844	1.6505	1.6505	1.6505	1.6505	1.6505	1.6505
λ_{min}	0.2837	0.3162	0.3181	0.3181	0.3181	0.3181	0.3181
$cond(K_J)$	5.9363	5.2190	5.1886	5.1885	5.1885	5.1885	5.1885

Table 2 Error e_J

J	e_J
5	5.534×10^{-1}
6	1.322×10^{-2}
8	2.853×10^{-3}
9	5.347×10^{-4}

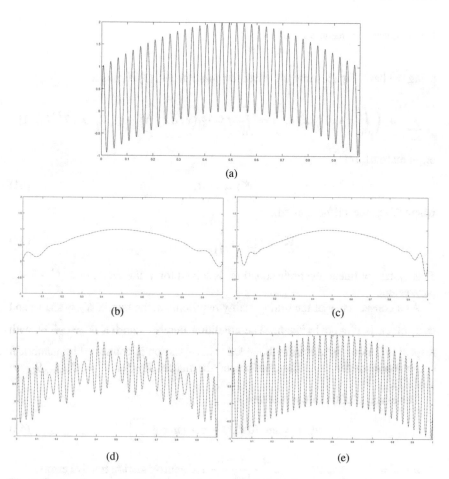

(a)

(b)　　　　　　　　　　　　　　　　(c)

(d)　　　　　　　　　　　　　　　　(e)

Fig. 7 Exact and approximate solutions u_J, $J = 3, 4, 5, 8$. (**a**) Exact solution. (**b**) Approximate solution for $J = 3$. (**c**) Approximate solution for $J = 4$. (**d**) Approximate solution for $J = 5$. (**e**) Approximate solution for $J = 8$

6   Conclusions

Due to the good properties of the proposed wavelet cubic B-splines basis, such as multiresolution analysis and the orthogonal characteristic according to inner product $\langle u', v' \rangle$, the numerical resolution of boundary value problems is easy and efficient. The matrix K_J involved in the linear system is block diagonal (each block is a banded matrix), and its condition number is bounded independently of the scale.

The work presented can be extended in several ways. The implemented technique using wavelet cubic B-splines bases could be well suited for solving nonlinear and higher-dimensional differential equations. We hope to address some of these problems in a future paper.

References

1. Chen, X., Yang, S., Zhengjia,J.: The construction of wavelet finite element and its application. Finite Elem. Anal. Des. **40**, 541–554 (2004)
2. Bindal, A., Khinast, J.G., Ierapetritou, M.G.: Adaptive multiscale solution of dymanical systems inchemical processes using wavelets. Comput. Chem. Eng. **27**, 131–142 (2003)
3. Cai, W., Wang, J.: Adaptive mutiresolution collocation methods for initial boundary value problems of nonlinear PDEs. SIAm J. Number. Anal. **33**, 937–970 (1996)
4. Kumar, V., Mehra, M.: Cubic spline adaptive wavelet scheme to solve singularly perturbed reaction diffusion problems. Int. J. Wavelets Multiresolution Inf. Process. **5**, 317–331 (2007)
5. Reddy, J.N.: On the numerical solution of differential equations by the finite element method. Indian J. Pure Appl. Math. 16 **12**, 1512–1528 (1985)
6. Jia, R.Q., Liu, S.T.: Wavelet bases of Hermite cubic splines on the interval. Adv. Comput. Math. **25**, 23–29 (2006)
7. Vampa, V., Martín, M.T., Serrano, E.: A hybrid method using wavelets for the numerical solution of boundary value problems on the interval. Appl. Math. Comput. **217**, 3355–3367 (2010)
8. Vampa, V., Martín, M.T., Serrano, E.: A new refinement Wavelet-Galerkin method in a spline local multiresolution analysis scheme for boundary value problems. Int. J. Wavelets Multirresolution Inf. Process. **11**, 1350015-1–1350015-19 (2013)
9. Černá, D., Finěk, V.: Wavelet basis of cubic splines on the interval on the hypercube satisfying homogeneous boundary conditions. Int. J. Wavelets Multiresolution Inf. Process. **13**, 1550014/1-21 (2015)
10. Han, B., Michelle, M.: Derivative-orthogonal Riesz wavelets in Sovolev spaces with applications to differential equations. Appl. Comput. Harmon. Anal. (2017). https://doi.org/10.1016/j.acha.2017.12.001
11. Ciarlet, P.G.: The Finite Element Method for Elliptic Problems. North Holland, Amsterdam/New York (1978)
12. Chui, C.K.: An Introduction to Wavelet Analysis. Academic, Boston (1992)
13. Meyer, Y.: Ondelettes sur l'intervalle. Rev. Mat. Iberoamericana **7**, 115–143 (1991)
14. Chui, C.K., Quak, E.: Wavelets on a bounded interval. In: Numerical Methods of Approximation Theory. International Series of Numerical Mathematics, pp. 53–75. Birkhäuser, Basel (1992)
15. Mallat, S.: A Wavelet Tour of Signal Processing – The Sparse Way. Academic/Elsevier MA EEUU. Burlington, MA (2009)

16. Walnut D.: An Introduction to Wavelet Analysis. Applied and Numerical Harmonic Analysis Series. Birkhäuser, Boston (2002)
17. Unser, M.: Ten good reasons for using spline wavelets. Proc. SPIE Wavelets Appl. Signal Image Process. V **3169**, 422–431 (1997)
18. Schoenberg, I.J.: Cardinal interpolation and spline functions. J. Aprox. Theory **2**, 167–206 (1969)
19. Primbs, M.: Stabile biorthogonale Spline-Waveletbasen auf dem Intervall. Dissertation, Universitat Duisburg-Essen (2006)

Kalman-Wavelet Combined Filtering

Guillermo La Mura, Ricardo Sirne, and Marcela A. Fabio

Abstract In this chapter we propose to combine two well-known techniques suitable for the analysis of nonstationary process: Kalman filtering and discrete wavelet transform. Signal filtering is an inverse problem in the sense that, based on the noisy observations obtained from measurements, it intends to estimate the state variables knowing the model of the system and the statistical behavior of the intervening noises. This technique performs simultaneously estimation and decomposition of random signals through a filter bank based on the Kalman filtering approach using wavelets. The algorithm introduced in the following pages takes advantage of the relative benefits of both methods. We present some numerical results considering two special cases of wavelets: Haar and Daubechies of four coefficients.

1 Introduction

In the last decades, many variations on the ensemble Kalman filtering have been published in the literature considering the problem of signal noise reduction and filtering techniques [1–3, 5, 13, 14, 16–19, 25]. Recently, schemes combining Kalman filtering and discrete wavelet transform have appeared [15, 21, 23, 24].

The main purpose of this work is to describe a filtering scheme for a signal with a certain type of additive noise, by combining the optimal iterative Kalman filtering process and the well-known technique based on discrete wavelet transform: wavelet denoising, on a multiresolution analysis (MRA) context. This technique performs

G. La Mura (✉) · M. A. Fabio
Centro de Matemática Aplicada, Universidad Nacional de San Martín, Buenos Aires, Argentina
e-mail: glamura@unsam.edu.ar; mfabio@unsam.edu.ar

R. Sirne
Facultad de Ingeniería, Departamento de Matemática, Universidad de Buenos Aires, Buenos Aires, Argentina
e-mail: rsirne@fi.uba.ar

© The Author(s), under exclusive license to Springer Nature Switzerland AG 2021
J. P. Muszkats et al. (eds.), *Applications of Wavelet Multiresolution Analysis*,
SEMA SIMAI Springer Series 4, https://doi.org/10.1007/978-3-030-61713-4_3

simultaneously estimation and decomposition of random signals through a filter bank based on the Kalman filtering approach using wavelets.

This chapter is organized as follows: in the next section we briefly review the classical Kalman filter, how does the filter work, and the optimal state estimation algorithm. Wavelets on a multiresolution analysis, the approximation scheme, and denoising by thresholds in the detail coefficients before reconstruction are introduced in Sect. 3. In Sect. 4 we developed the ensemble Kalman-wavelet filtering scheme. Some numerical examples are presented in Sect. 4.1. Finally, we conclude in Sect. 5 with a discussion of other potential applications of this technique as well as future improvements.

2 Kalman Filters

2.1 What Are Kalman Filters, and How Do They Work?

The Kalman filter is an efficient recursive filter that estimates the state of a linear dynamic system from a series of noisy measurements. It is an important topic in control theory that it is used in a wide range of engineering applications: from radar, biosignals, to estimation of macroeconomic models.

The Kalman filter is an optimal estimator, knowing the dynamic system model (e.g., physical laws of motion), the control input variables and multiple sequential measurements (usually observed through sensors), this method allows to predict the state of the system and updating estimation at each measurement.

First, let us consider the linear dynamical systems without random perturbations by the mathematical model

$$\begin{cases} \dot{x}(t) = Ax(t) + Bu(t), \\ y(t) = Cx(t) \end{cases} \tag{1}$$

represented in the first block of Fig. 1, where A, B, and C are known constant matrices.

Generally, the state variable $x(t)$ is not observable; we can only measure the output $y(t)$. Since the system model is known, the state and its output can be estimated, and we denote these estimated variables as $\hat{x}$ and $\hat{y}$, respectively. The observation error ε_{ob} and the state error ε_{st} are obtained as the difference between the system variables and its estimations. We can observe that in the state function the correction factor is the gain of Kalman (K) times the measurement or observation error:

$$\dot{\varepsilon}_{st} = (A - KC) \cdot \varepsilon_{st} \tag{2}$$

Fig. 1 Linear system and
mathematical model.
Real-time Kalman filtering

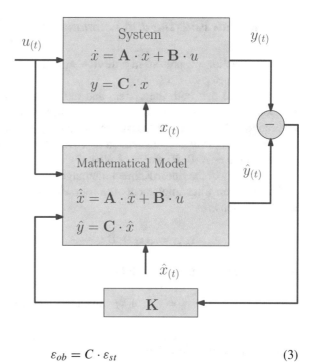

$$\varepsilon_{ob} = C \cdot \varepsilon_{st} \tag{3}$$

$$\dot{x} = Ax + Bu \qquad\qquad\qquad y = Cx$$

$$\dot{\hat{x}} = A\hat{x} + Bu + K(y - \hat{y}) \qquad \hat{y} = C\hat{x}$$

$$\underbrace{\dot{x} - \dot{\hat{x}}}_{\dot{\varepsilon}_{st}} = Ax - A\hat{x} - K(y - \hat{y}) \qquad \underbrace{y - \hat{y}}_{\varepsilon_{ob}} = C\underbrace{(x - \hat{x})}_{\varepsilon_{st}}$$

Returning to the mathematical model (1), we see how the observation error is modulated by Kalman's gain, ensuring an optimal approximation for the iterated update of the estimated variables. Note that the errors are modelled by a system of equations that are similar to those of the original system (see (2) and (3)). If the model and the measurement are contaminated with additive noise, these perturbations will propagate to both the observed and the state.

According to (2), when the real part of the eigenvalues of $(A - KC)$ is less than zero, the slope of the state error decreases, that is, the observation error tends to zero when time tends to infinity. In Sect. 2.3 we will show an example of how the error decreases as time increases.

The estimation of these errors and their covariances enable us to update the estimate, but to do it we need three variables: the input control signal $u(t)$, the measured signal $y(t)$, and the prediction of the estimated state $\hat{x}(t)$.

2.2 *Kalman Filtering and Estimation Algorithm*

In this subsection we present in a discrete approach to the system, the model, and the time, with step k.

We assume that both the system and the measurement are contaminated with independent random noise of Gaussian distribution with zero mean, w and v, respectively. Moreover, the measurement noise v_k is uncorrelated with the process noise w_k. Then, if w is characterized by covariance Q and v is by covariance R, it follows that $v \sim N(0, R)$ and $w \sim N(0, Q)$.

A summary of the linear Kalman filter algorithm is shown in Table 1.

The Kalman filter implements the following linear discrete-time process with state, x, at step k:

$$x_{k+1} = Ax_k + Bu_k + w_k \quad \text{(state equation)}, \tag{4}$$

and the measurement, y, is given by

$$y_k = Cx_k + v_k \quad \text{(measurement equation)}, \tag{5}$$

Usually, $x, w \in \mathbb{R}^{n \times 1}, u \in \mathbb{R}^{p \times 1}, v \in \mathbb{R}^{m \times 1}, A \in \mathbb{R}^{n \times n}, B \in \mathbb{R}^{n \times p}$, and $C \in \mathbb{R}^{m \times n}$.

The first block of Fig. 2 represents the linear system with noise, defined in (4) and (5).

Table 1 Kalman filter algorithm

Step 1	Filter initialization	Initialize $\hat{x}_0$ and $\hat{P}_0$
Step 2	State prediction	$\begin{cases} \hat{x}_k^- = A\hat{x}_{k-1}^- + Bu_k \\ \hat{P}_k^- = AP_{k-1}A^T + Q \end{cases}$
Step 3	Observation-related prediction	$y_k^- = C\hat{x}_k^-$
Step 4	Kalman filter update	$\begin{cases} K_k = P_k^- C^T (CP_k^- C^T + R)^{-1} \\ \hat{x}_k = \hat{x}_k^- + K_k(y_k - y_k^-) \\ P_k = (I - K_kC)P_k^- \end{cases}$
Step 5	Store results	Store $\hat{x}_k$ and $\hat{P}_k$
Step 6	Return to Step 2	

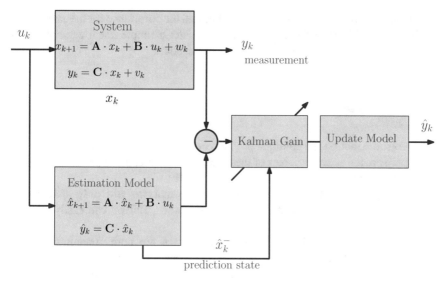

Fig. 2 Discrete time Kalman filtering

The Kalman filter "prediction-update" algorithm computes the following two stages recursively:

1. Prediction: Process parameters $\hat{x}_k$ (state) and $\hat{P}_k$ (state error covariance) are estimated using the a priori state $\hat{x}_k^-$ and $\hat{P}_k^-$:

$$\hat{x}_k = \underbrace{A\hat{x}_{k-1} + Bu_k}_{\hat{x}_k^-} + K_k(y_k - C\underbrace{(A\hat{x}_{k-1} + Bu_k)}_{\hat{x}_k^-})$$

2. Update: The state and error covariance are corrected using the current measurement.

The first stage uses the previously estimated state, $\hat{x}_{k-1}^-$, to predict the current state in the step k, $\hat{x}_k^-$, as shown the following equation:

$$\begin{cases} \hat{x}_k^- = A\hat{x}_{k-1}^- + Bu_k \\ \hat{P}_k^- = AP_{k-1}A^T + Q \end{cases} \qquad (6)$$

In the second stage, using the current measurement, y_k, and the predicted state, x_k, it is estimated the current state value at step k, $\hat{x}_k$, and is obtained a more accurate approximation:

$$\begin{cases} K_k = P_k^- C^T (C P_k^- C^T + R)^{-1} \\ \hat{x}_k = \hat{x}_k^- + K_k(y_k - C\hat{x}_k^-) \\ P_k = (I - K_k C)P_k^- \end{cases} \qquad (7)$$

updating the Kalman gain, the estimated state and the covariance for the same step k.

In order to illustrate these ideas, let us consider a simple linear example of the movement of a projectile, before we close this section.

2.3   Example

The projectile motion is a common problem in physics, where acceleration, velocity, and space are the states variables of the model. In this case, the observed dimensions of the system are two: displacement and height [12].

To simplify the state equations, we assume that we will only analyze two variables: distance and altitude. The model is known, and we can easily simulate it.

The measurements of the altitude and distance are both contaminated with Gaussian noise. We can apply Kalman filtering process and observe how quickly the estimated values approach the theoretical trajectory. As it arises from the previous explanation, the estimation error is significant, and the tracking of the trajectory is not very good; see Fig. 3, (right). However, the error decreases as time grows; see Fig. 3, (left).

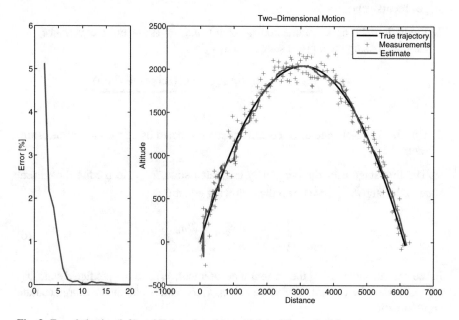

Fig. 3 Error behavior (left) and Kalman's estimate (right) of Example 2.3

3 Wavelets on a Multiresolution Analysis Scheme

In this section we present an introduction to multiresolution signal approximation.

Let $L^2(\mathbb{R})$ denote the space of real valued functions with finite energy in the real time domain, in which the inner product is defined by $< f, g >= \int_{\mathbb{R}} f(t)g(t)dt$ and the norm is defined by $||f||_{L^2} = \sqrt{|< f, f >|}$.

A wavelet well localized in time and frequency is an oscillating function with zero mean, which decays rapidly in time. From one "mother wavelet" ψ a two-parameter family of wavelets ψ_{ab} is obtained by translations in the time variable and dilations by a scale factor, with influences the location of the time-frequency window and the width of the corresponding time and frequency windows. This family is

$$\left\{\psi_{ab}(t) = |a|^{-1/2}\, \psi(\frac{t-b}{a}), \ a, b \in \mathbb{R}, a \neq 0\right\}.$$

It is computationally convenient to discretize the latter by restricting the analysis to the $a = 2^j$ scales with $b = k2^j$ translations, adopting the following nomenclature:

$$\left\{\psi_{jk}(t) = 2^{-j/2}\, \psi(2^{-j}t - k), \ j, k \in \mathbb{Z}\right\}.$$

This family of functions is constructed with the objective of expressing $s(t)$ signal with the following serial development of wavelets:

$$s(t) = \sum_{j \in \mathbb{Z}} \underbrace{\sum_{k \in \mathbb{Z}} d_j(k)\psi_{jk}(t),}_{r_j(t)}$$

where convergence in 2-norm is assumed. Particularly, r_j is called the j-th residual signal. Each $r_j(t)$ shows the behavior of $s(t)$ in the j-th frequency band, centered on the frequency f_j, fulfilling $f_j/f_{j+1} = 2$. Thus, consecutive band centers are separated by an octave, and the discretization is called dyadic (Fig. 6).

The implementation of the aforementioned decomposition can be carried out efficiently in the framework of a multiresolution analysis (MRA) [4, 6, 7, 22].

A MRA is a nested sequence V_j of closed subspaces of $L^2(\mathbb{R})$, such that

1. $V_{j+1} \subset V_j, \ j \in \mathbb{Z}$
2. $\bigcap_{j \in \mathbb{Z}} V_j = \{0\}$
3. $\bigcup_{j \in \mathbb{Z}} V_j$ is dense in $L^2(\mathbb{R})$
4. $s(t) \in V_0 \Leftrightarrow s(2^{-j}t) \in V_j, \ j \in \mathbb{Z}$
5. $s(t) \in V_0 \Leftrightarrow s(t - k) \in V_0, \ k \in \mathbb{Z}$
6. exist a scale function ϕ such that $\{\phi(t - k), k \in \mathbb{Z}\}$ is a Riesz basis of V_0.

Following the previous notation, the scale function ψ generates

$$\{\phi_{jk}(t) = 2^{-j/2}\,\phi(2^{-j}t - k), \ j, k \in \mathbb{Z}\}.$$

Since $j > 0, 2^j \in \mathbb{Z}$, it results:

$$s(t) \in V_0 \underbrace{\Leftrightarrow}_{(5)} s(t - 2^j k) \in V_0 \underbrace{\Leftrightarrow}_{(4)} s(2^{-j}t - k) \in V_j.$$

Consequently, given $\phi \in V_0$, the function $\phi_{jk} \in V_j$.

It is said that wavelet ψ is orthogonal when the inner products satisfy

$$< \psi_{jk}, \psi_{mn} >= \delta_{jm}, \delta_{kn}, \ j, k \in \mathbb{Z},$$

where δ_{ij} is the Kroneker delta.

In this context, if

$$W_j = \text{span}\{\psi_{jk}, k \in \mathbb{Z}\}$$

$L^2(\mathbb{R})$ can be expressed as

$$L^2(\mathbb{R}) = \cdots \oplus W_{-1} \oplus W_0 \oplus W_1 \oplus \cdots$$

Each $s \in L^2(\mathbb{R})$ has a unique decomposition

$$s(t) = \cdots + r_{-1}(t) + r_0(t) + r_1(t) \cdots,$$

where each $r_j \in W_j$ is expressed as a linear combination of the base $\{\psi_{jk}, k \in \mathbb{Z}\}$. Since $V_j = W_{j+1} \oplus W_{j+2} \oplus \cdots$ and $V_j = W_{j+1} \oplus V_{j+1}$, with $j \in \mathbb{Z}$, if $s_0 \in V_0$, it can be expressed as

$$s_0(t) = s_J(t) + \sum_{j=1}^{J} r_j(t), \tag{8}$$

where

$$r_j(t) = \sum_{k \in \mathbb{Z}} d_j(k)\psi_{jk}(t) \ \text{and} \ s_J(t) = \sum_{k \in \mathbb{Z}} c_J(k)\phi_{Jk}(t). \tag{9}$$

Being s_0 a band-limited signal, J can be chosen such that all the information of interest is concentrated in the r_j $j = 1, \cdots, J$, leaving in s_J an insignificant part of the energy of the signal with the lower frequency composition. This allows to approximate s_0 as

$$s_0(t) \approx \sum_{j=1}^{J} r_j(t). \tag{10}$$

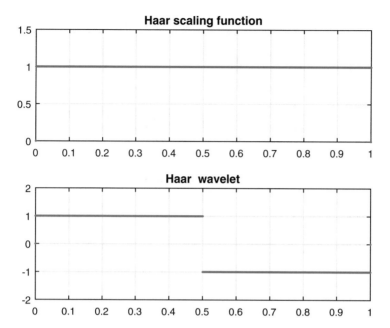

Fig. 4 Haar wavelet and its scale function

In this framework, an efficient analysis methodology to approximate $s_0(t)$ is achieved by selecting J and calculating the coefficients $d_j(k)$ accurately to approximate $s_0(t)$, see (8).

Indeed, the MRA structure allows fast and exact calculation of the wavelet coefficients of a $L^2(\mathbb{R})$ signal by providing a recursion relationship between the scaling coefficients at a given scale and the scaling and wavelet coefficients at the next coarser scale.

In this work we choose Haar (Fig. 4) and Daubechies (Fig. 5) wavelets and scaling functions [4, 6, 7].

3.1 The Haar and Daubechies Wavelets

Using the Haar wavelet, the coefficients can be calculated recursively by

$$c_j(k) = \frac{1}{\sqrt{2}}(c_{j-1}(2k)+c_{j-1}(2k+1)) \text{ and } d_j(k) = \frac{1}{\sqrt{2}}(c_{j-1}(2k)-c_{j-1}(2k+1))$$

$$(11)$$

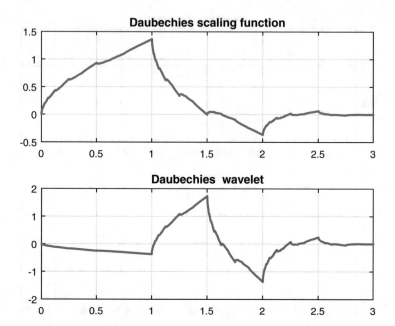

Fig. 5 Daub-4 wavelet and its scale function

whereas using the wavelet of Daubechies with four coefficients (Daub-4), the decomposition coefficients are recursively obtained as

$$c_j(k) = \sum_{i=0}^{3} p_i c_{j-1}(2k+i) \quad \text{and} \quad d_j(k) = \sum_{i=-2}^{1} (-1)^i p_{1-i} c_{j-1}(2k+i) \quad (12)$$

with $\quad p_0 = \dfrac{1+\sqrt{3}}{4\sqrt{2}}, p_1 = \dfrac{3+\sqrt{3}}{4\sqrt{2}}, p_2 = \dfrac{3-\sqrt{3}}{4\sqrt{2}}$, and $p_3 = \dfrac{1-\sqrt{3}}{4\sqrt{2}}$.

Other implementation details are indicated in the following sections (Fig. 6).

3.2   MRA Algorithm

In practice, the MRA of a discrete signal $s(n)$ is performed by successive application of the two-channel filter bank: low-pass and high-pass filters. In this scheme, these two filters H_0 and H_1 decompose the signals into low-pass and high-pass components subsampled by 2 (analysis process) and perform the decomposition (8); see Fig. 6. In signal processing, H_0 and H_1 are called quadrature mirror filters.

In this step, if necessary, the noise elimination process can be performed.

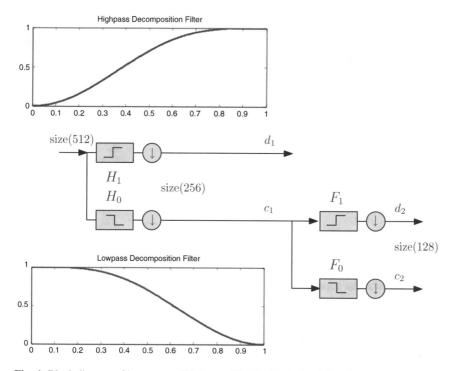

Fig. 6 Block diagram of low-pass and high-pass filter for Daub-4 and $J = 2$

The original signal can be recovered from the application of two filters F_0 and F_1 and upsampling (synthesis process).

Filters must meet some constraints in order to produce a perfect reconstruction of the signal.

Next, we list the considerations to implement the AMR algorithm:

- The signal is analyzed in blocks of 2^N values, corresponding 2^{J-j} coefficients of each type at resolution level $1 \leq j \leq J$.
- If the signal s_0 is contaminated with additive noise v_k with null mean and deviation σ, supposing that v_k and v_{k+1} are uncorrelated, the coefficients of the MRA inherit additive noise with equal mean and deviation. In particular, if the noise is Gaussian, the coefficients are also Gaussian.
- If the signal has a frequency content in some interval $[f_{min}, f_{max}]$, the MRA is performed for the levels $j = 1, \cdots, J$ such that s_J have no significant energy. Then, it can be considered

$$s_0 \approx r_1 + \cdots + r_J.$$

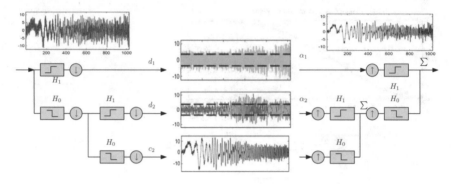

Fig. 7 Thresholding example

3.3 Denoising by Thresholds in the Wavelet Detail Coefficients

As it is known, the general noise elimination procedure involves three steps:

1. Decomposition: compute the wavelet decomposition of the signal at a chosen J.
2. Threshold detail coefficients: for each level from 1 to J, select a threshold and apply soft thresholding to the detail coefficients.
3. Reconstruction: compute wavelet reconstruction using the original approximation coefficients of level J and the modified detail coefficients of levels from 1 to J.

The second step is very important because the threshold value is applied to remove noise by thresholding function:

$$\tilde{d}_j(k) = \begin{cases} sg(d_j(k))L(|d_j(k)|), & \text{if } |d_j(k)| > \alpha_j \\ 0 & \text{if } |d_j(k)| < \alpha_j \end{cases}$$

where L is usually a linear function and α_j is a threshold for the resolution level j. Then, the selections of the threshold value affect the quality of signal.

The main problem with this method is to find the optimal threshold α_j, [8–11].

This filtering scheme is efficient when the signal energy is concentrated in some resolution levels; see Fig. 7.

4   Kalman-Wavelet Combined Filtering

A way to apply these computational schemes jointly is shown in the block diagram appearing in Fig. 8. Unfortunately it has high computational cost. Additionally, the Kalman filter does not optimize the signal-to-noise ratio for each resolution level.

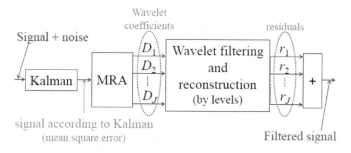

Fig. 8 Signal processing by first applying Kalman and then wavelet

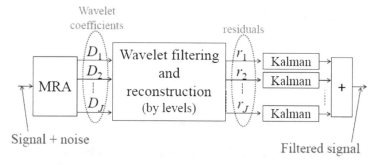

Fig. 9 Signal processing by first applying wavelet and then Kalman

On the other hand, if we apply them in the opposite order, according to the block diagram exposed in Fig. 9, Kalman's hypothesis for residuals is not verified, and the procedure has high computational cost too.

We will introduce an effective technique of Kalman-wavelet filtering by simultaneous estimation and decomposition of random signals through a filter bank based on Kalman filtering approach using wavelets.

Since Kalman and wavelet filtering are radically different, it is possible to combine both with the objective of enhancing its advantages for certain types of signals, corresponding to linear systems modelled as indicated in the equations (4) and (5).

Starting from $c_0 = s_0$, i. e., considering the values of the sampled signal as coefficients at level $j = 0$, if s_0 admits Kalman filtering, then according to (11) or (12) the successive c_j and d_j will admit Kalman filtering too (developed in [20] for Haar case). Under the conditions described above and assuming that s_0 is contaminated with additive Gaussian noise of zero mean and deviation σ, it follows that c_j and d_j have the same type of noise.

The proposed filtering scheme consists in applying Kalman process to the coefficients d_j obtaining the d_j^K, and then, if necessary, filtering is carried out as the described Sect. 3.3 obtaining the $\tilde{d}_j^K$. The final filtered signal is achieved by (9) with the $\tilde{d}_j^K$, considering decomposition (8). This scheme is represented in Fig. 10.

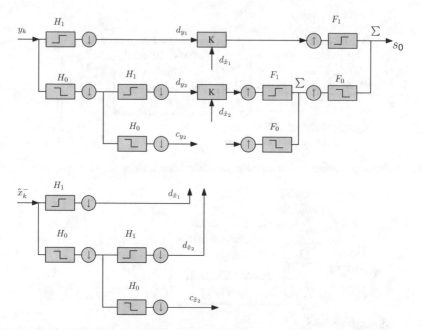

Fig. 10 Block diagram of combined Kalman-wavelet filter algorithm for $J = 2$

Below we consider a toy example of this filtering technique.

4.1   Example: Kalman-Wavelet Combined Filtering

To illustrate the application of the ensemble Kalman-wavelet filtering algorithm, we consider a simulated linear system that responds to the following continuous time model:

$$\begin{cases} \dot{x}(t) = A \cdot x(t) + B \cdot u(t) \\ y(t) = C \cdot x(t) + v(t) \end{cases} \tag{13}$$

where

$$A = \begin{bmatrix} 0 & 1 \\ -5 & -4 \end{bmatrix}, \; B = \begin{bmatrix} 1 & 0 \\ 0 & 1 \end{bmatrix}, \; C = \begin{bmatrix} 1 & 1 \end{bmatrix}$$

$$x_0 = \begin{bmatrix} 0 \\ 0 \end{bmatrix}, \quad u(t) = \begin{bmatrix} 0 \\ f(t) \end{bmatrix}$$

$$f(t) = 100 \sin(4\pi t) + 800 \cos(16\pi t)$$

$$v(t) \sim N(0, R) \; \text{Gaussian Noise}$$

System (13) in its pure state, i.e., without noise, is modeled by the following initial value problem:

$$\begin{cases} z'' + z' + 2z = f(t) \\ z(0) = z'(0) = 0 \end{cases} \qquad (14)$$

We can see in Fig. 11 that the Kalman-wavelet combined filtering sticks to the pure signal more accurately.

A local view of the first 2.5 s appears in Fig. 11. The first curve is the pure signal, after contaminated with additive noise, then filtered with Kalman and finally Kalman wavelet combined in two variants: Kalman-Daub-4 and Kalman-Haar. Except for the portion marked with red dotted line on the third curve, these techniques do not show significant difference with the original signal.

Table 2 shows the difference in signal-to-noise (S/N) levels achieved for two maximum wavelet resolution levels. The S/N values did not show significant differences in different simulations of the specified noise.

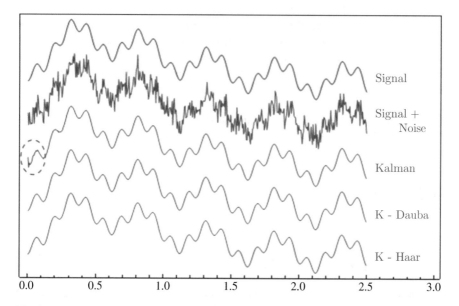

Fig. 11 Kalman-wavelet combined filtering performance

Table 2 Comparison S/N ratio

	$J = 10$	$J = 12$
Signal + Noise	27 dB	26 dB
Kalman filtering	82 dB	86 dB
Kalman - Haar	102 dB	112 dB
Kalman - Dauda 4	114 dB	122 dB

Given the structure of the AMR scheme, it allow us to decompose the signal by a bank of perfect reconstruction filters. It is observed that, in this context, the application of the Kalman filter on the wavelet coefficients of each resolution level improves the ratio S/N of the filtered signal compared to the direct application of the Kalman filter to the noisy signal, in particular with Daub-4 versus Haar.

5 Conclusions and Future Work

In this work we have proposed the extension of the Kalman-wavelet technique for the filtering of certain types of noisy signals. We explained the algorithm that combines the advantages of the Kalman recursive method with the flexibility of the hierarchical decomposition of approximations and details of a signal, provided by wavelet processing in the context of a multiresolution analysis.

We discuss in an example that the proposed method exhibits better performance in contrast to the direct application of Kalman filtering, for signals with concentrated frequency composition (in some resolution levels).

In many applications, the linearity requirement of the Kalman filter is not enough to model the dynamics system or the observation signal. Extensions and generalizations to the method have also been developed, such as the extended Kalman filter (EKF) and the unscented Kalman filter (UKF). The combination of EKF/UKF and discrete wavelet transform is being studied.

Acknowledgments The authors are especially grateful to Eduardo P. Serrano for sharing his ideas on the combination of these methods that give rise to this work.

References

1. Anderson, J.L.: An ensemble adjustment Kalman filter for data assimilation. Mon. Weather Rev. **129**, 2884–2903 (2001)
2. Anderson, J.L., Anderson, S.L.: A Monte Carlo implementation of the nonlinear filtering problem to produce ensemble assimilations and forecasts. Mon. Weather Rev. **127**, 2741–2758 (1999)
3. Bhaumik, S.: Robust ensemble kalman filter based on exponential cost function. Asian J. Control **16**(5), 1522–1531 (2014)
4. Chui, C.K.: An Introduction to Wavelets. Academic, Inc (1992)
5. Chui, C.K., Chen, G.: Kalman Filtering, 5th Edn. Springer (2017)
6. Daubechies, I.: Orthonormal Bases of Compactly Supported Wavelets, Communications on Pure and Applied Mathematics, XLI, pp. 909–996. Wiley, Inc (1988)
7. Daubechies, I.: Ten Lectures on Wavelets. SIAM (1992)
8. Donoho, D.L., Johnstone, I.M.: Ideal spatial adaptation by wavelet shrinkage. Biometrika **81**, 425–455 (1994)
9. Donoho, D.L.: Denoising by soft-thresholding. IEEE Trans. Inf. Theory **1**, 613–627 (1995)
10. Donoho, D.L., Johnstone, I.M.: Adapting to unknown smoothness via wavelet shrinkage. J. Am. Statist. Assoc. **90**(432), 1200–1224 (1995)

11. Donoho, D.L.: De-noising by soft thresholding. IEEE Trans. Info. Theory **43**, 933–936 (1993)
12. Giron-Sierra, J.M.: Digital Signal Processing with Matlab Examples, vol. 3. Springer (2017)
13. Hamill, T.M., Whitaker, J.S., Snyder, C.: Distance-dependent filtering of background error covariance estimates in an ensemble Kalman filter. Mon. Weather Rev. **129**, 2776–2790 (2001)
14. Hunt, B.R., Kostelich, E., Szunyogh, I.: Efficient data assimilation for spatiotemporal chaos. A local ensemble transform Kalman filter. Physica D Nonlinear Phenom. **230**, 112–126 (2007)
15. Hickmann, K.S., Godinez, H.C.: A multiresolution ensemble Kalman filter using the wavelet decomposition. Comput. Geosci. **21**, 441–458 (2017). https://doi.org/10.1007/s10596-017-622-7
16. Houtekamer, P.L., Mitchell, H.L.: Data assimilation using an ensemble Kalman filter technique. Mon. Weather Rev. **126**, 796–811 (1998)
17. Kalman, R.E.: A new approach to linear filtering and prediction problems. Trans. ASME J. Basic Eng. **82**, 35–45 (1960)
18. Kalman, R.E., Bucy, R.S.: New results in linear filtering and prediction theory. Trans. ASME J. Basic Eng. **82**, 95–108 (1961)
19. Keppenne, C.L.: Data assimilation into a primitive-equation model with a parallel ensemble Kalman filter. Mon. Weather Rev. **128**, 1971–1981 (2000)
20. La Mura, G., Sirne, R.O., Serrano, E.P.: Algoritmo conjunto Kalman-Wavelet para el fltrado de señales, [in Spanish], Actas MACI (2011)
21. La Mura, G., Sirne, R.O., Fabio, M.A.: Algoritmo conjunto Kalman-Wavelet para el fltrado de señales (II), [in Spanish], Actas MACI (2019). ISSN: 2314-3282
22. Mallat, S.G.: A theory for multiresolution signal decomposition: the wavelet representation. IEEE Trans. Patt. Anal. Machine Intell. **2**(7), 674–693 (1989)
23. Shahtalebi, S., et.al.: WAKE: Wavelet Decomposition Coupled with Adaptive Kalman Filtering for Pathological Tremor Extraction. arXiv:1711.06815v2 [eess.SP] (2018)
24. Viegener, A., Serrano, E., et al.: Algoritmo conjunto Kalman-Haar aplicado al procesamiento de señales. Revista de Matemática: Teoría y Aplicaciones **19**(1), 37–47 (2012)
25. Whitaker, J.S., Hamill, T.M.: Ensemble data assimilation without perturbed observations. Mon. Weather Rev. **130**, 1913–1924 (2002)

Using the Wavelet Transform for Time Series Analysis

María B. Arouxet, Verónica E. Pastor, and Victoria Vampa

Abstract The characterization of time series requires knowledge of certain parameters. One of those parameters is the Hurst exponent, which is an indicator of long-range dependence characteristics. Rescaled range (R/S), proposed by E. Hurst, is the most commonly used method to compute this exponent. On the other hand, wavelet analysis is known to reflect better the nonlinear dynamics of the biological, climatic, or economic series than the statistical tools often used for this analysis. The average wavelet coefficient (AWC) is a wavelet method that has been used for the last years to compute the Hurst exponent. In this paper, we present a modification to the AWC method, and we compare its performance with the original version of AWC and with R/S methods. The results obtained for the synthetic series were so promising that we decided to apply our proposal in rainfall series. Therefore, these results were compared with the ones reported from La Pampa. After that, series from different climatic regions of the Argentine Republic were analyzed.

1 Introduction

The characterization of complex systems is not easy, since they cannot be split into simpler subsystems without losing dynamic properties. Sometimes, the systems of equations are not known, and only one observable of the systems is known. This observable is represented as a time series, which is a collection of observations

M. B. Arouxet
Facultad de Ciencias Exactas-CMALP, Universidad Nacional de La Plata, La Plata, Argentina
e-mail: belen@mate.unlp.edu.ar

V. E. Pastor
Facultad de Ingeniería, Universidad de Buenos Aires, CABA, Argentina
e-mail: vpastor@fi.uba.ar

V. Vampa (✉)
Facultad de Ingeniería, Departamento de Ciencias Básicas, Universidad Nacional de La Plata, La Plata, Argentina
e-mail: victoria.vampa@ing.unlp.edu.ar

© The Author(s), under exclusive license to Springer Nature Switzerland AG 2021
J. P. Muszkats et al. (eds.), *Applications of Wavelet Multiresolution Analysis*,
SEMA SIMAI Springer Series 4, https://doi.org/10.1007/978-3-030-61713-4_4

$\{X(t)\}$ obtained through repeated measurements over time. For example, we can consider series of mean monthly rainfalls recorded at scattered locations, daily coin values, or the RR intervals of electrocardiogram signals. They are time series of different nature measured in equally spaced intervals, different in each case. If the data is collected irregularly, it is not a time series.

As it is known, an observed time series can be decomposed into three components: the trend (T: long term direction), the seasonality (S: systematic, calendar-related movements), and the irregularity (I: unsystematic, short-term fluctuations),

$$X(t) = T(t) + S(t) + I(t) \tag{1}$$

and different techniques have been developed to study these components.

Long-range dependence (LRD) is the dependence between observations far away in time and has attracted strong interest in recent years [1, 2]. Its utility has been analyzed in hydrology [3], in economics [4], in finance [2], and in many other fields. In signals that exhibit long-range dependent features, correlations persist on very long time scales and also have a certain self-similar structure. A natural way to analyze LRD is performing a rescaling operation and studying how measured properties vary as a function of the scale.

One of the techniques used to measure LRD is the Hurst exponent which was introduced by E. Hurst in 1951 [5] when he studied the fluctuations of the water level in the Nile River. Hurst found that a high water level in the river flow means a wonderful harvest. He discovered a trend after many years of this studies: years of abundant rain were continued of years with high level of water in the river flow; years of drought were continued with years with the same trend. The Hurst exponent, H, is a real parameter that varies between 0 and 1 and indicates the persistence level of a process. Values greater than (less than) 0.5 are associated to persistent (anti-persistent) processes, and $H = 0.5$ can be seen as uncorrelated white noise.

Since processes with LRD are self-affine processes [6], the techniques used to measure LRD are the same to those to find self-affinity. Most popular examples of self-affine processes are Brownian motion and Gaussian white noise. Both processes can be generalized by introducing the concept of fractional differentiation leading to fractional Brownian motion (fBm) and fractional Gaussian noise (fGn).

Many methods to estimate H are used under the assumption that the series studied are fBm-type (or fGn-type), so, as is discussed in Serinaldi [2], it is very important to distinguish between fBm and fGn type series in order to estimate H exponent correctly.

We focused our research on the study of estimating Hurst exponent within the wavelet transform framework and proposing a modification of the AWC method [7], using a more robust variance estimator of detail coefficients [8, 9]. We analyze two classes of time series: synthetic series provided by computational algorithms with predefined Hurst exponents and rainfall series from different regions in Argentina, presenting climatological variability. Rainfall prediction is a very challenging task due to its dependence on many meteorological parameters. Because of the complex nature of rainfall, the uncertainty associated with its predictability continues to be an issue in rainfall forecasting. In the literature there are few reports of Hurst exponents in rainfall series from South America [3, 10], only one of them is from Argentina [11].

The rest of the paper is divided into five sections. In the next section methods for estimating Hurst exponent are reviewed. The new proposal is also presented. Section 3 provides a description of time series we analyzed, and in Sect. 4 a comparison of the results that were obtained in each case is shown. In Sect. 5 we give some concluding remarks, and directions for future research are mentioned.

2   H Estimators

There are different ways to analyze signals which exhibit long-range-dependent features. They have been used as a basis for estimation of the Hurst exponent H. In this work it is shown how an estimator based on the wavelet transform and in a multiresolution analysis framework can be defined. The proposed estimator is related to the details rather than the approximation and has good statistical and computational properties. It is not necessary to detect the type of signal (stationary or nonstationary) before applying this H estimator. This is a good advantage with respect to other techniques where a prior knowledge of the type signal is required. Some analysis about the anti-persistence of some series is shown to be incorrect because of the inappropriate use of some estimation methods. Research supports the opinion that methods designed for stationary fGn can fail to provide correct values of H if they are applied to nonstationary fBm. A theoretical discussion about this point can be found in Serinaldi, [2], and in Cannon et.al., [12].

2.1   The Rescaled Range (R/S)

This method to compute Hurst exponent [5, 13] has been used for many authors and is analyzed in many articles [2, 14].

Given a time series $x_n, n = 1, \ldots, N$, the method comprises the following steps:

- For $n = 1, 2, \ldots, N$

  - The mean, $m = \frac{1}{n} \sum_{i=1}^{n} x_i$, and the summation of time series relative to m, $y_n = x_n - m$, i.e., the cumulative deviate series z: $z_n = \sum_{i=1}^{n} y_i$ are calculated.
  - Range series, $R_n = \max(z_1, z_2, \ldots, z_n) - \min(z_1, z_2, \ldots, z_n)$
    and a standard deviation of the series, $S_n = \sqrt{\frac{1}{n} \sum_{i=1}^{n} (x_i - m)^2}$,
    are created.
  - The rescaled range statistics series (R/S)
    $(R/S)_n = \frac{R_n}{S_n}$ is calculated.

The foundation of the method is that, based on the self-affinity, it can be expected that $(R/S)_n \approx kn^H$, i.e., $E\left[(R/S)_n\right] \approx kn^H$, for $n \to \infty$, where $E[x]$ is the expected value and k is a positive constant that is not dependent on n.

The Hurst exponent is derived by plotting $(R/S)_n$ as a function of $log(n)$ and fitting a straight line. The slope of the line gives H estimator.

2.2   Wavelet Analysis

In this section we give a brief sketch of the multiresolution analysis aspect of wavelets and also about the continuous and the discrete wavelet transforms.

As is described with detail in [15–17], a multiresolution analysis (MRA) consists in a collection of nested subspaces V_j, $j \in Z$ satisfying a set of properties. MRA involves successively projecting a signal $f(s)$ to be studied into each of the *approximations subspaces* V_j. Since $V_j \subset V_{j+1}$, the approximation in V_j is coarser than the approximation in V_{j+1}. The information that is removed when going from one approximation to the next coarser is called the detail, and MRA shows that the detail signals can be obtained from the projections of f onto a collection of subspaces, the W_j, called the *wavelet subspaces*. In other words, the information in a signal f can be written as a collection of details at different resolutions or scales and a low-resolution or coarse approximation.

From MRA theory there exists a function ψ, called the mother wavelet, derived from ϕ, the scale function, such that its templates constitute a Riesz basis for W_j. As the approximations in V_j are coarser and coarser approximations of f, ϕ needs to be a low-pass filter; on the other hand, ψ is a band-pass function.

The continuous wavelet transform (CWT) is

$$W_\psi f(a, b) = \int_{-\infty}^{\infty} f(t)\psi_{a,b}(t)dt \qquad a, b \in R, a > 0 \tag{2}$$

where $\psi_{a,b}(t) = \frac{1}{\sqrt{(a)}}\psi(\frac{t-b}{a})$ are basis functions obtained by scaling and shifts of a mother wavelet function $\psi(t)$, which must satisfy $\int_{-\infty}^{\infty} \psi(t)dt = 0$ and its Fourier transform $\hat{\psi}(\omega)$ is a band-pass filter, with fast decay. More details about the admissibility condition can be found in [15]. The dilation and translation parameters, a and b, respectively, vary continuously. The transform $W_\psi f(a, b)$ can be interpreted as the energy of f of scale a at $t = b$ and is often represented graphically and plotted as two-dimensional images.

If the mother wavelet is real and the signal $f(t)$ has finite energy, the discrete wavelet transform can be written as

$$DW_\psi f(j, k) = \langle f, \psi_{j,k} \rangle$$

where j and $k \in Z$ are, in general, dyadic values of a and b. In this way the DWT can be seen as a mapping from $L^2(R) \rightarrow l^2(Z)$ given by

$$f(t) \rightarrow \{\{a(J, k), k \in Z\}, \{d(j, k), j = 1, \cdots, J, k \in Z\}\}.$$

These coefficients are defined through inner products of f with shifted and scaled versions of the scaling function ϕ and the mother wavelet ψ, whose definition depends on whether one chooses to use an orthogonal, semi-orthogonal, or bi-orthogonal DWT [18]. They are computed in practice from a fast recursive algorithm which has low computational cost.

2.2.1 The Average Wavelet Coefficient Method

A time series $X(t)$ is called self-similar with self-similarity parameter H (or H-self similar), if for any positive scale factor c satisfies a power law $X(ct) \approx c^H X(t)$. In this case, a scale parameter will affect assimptotically the variance of wavelet transform, hereinafter referred to as $W(a, b)$ (Eq. 2):

$$Var(a) = \mathbb{E}(W(a, b))^2 - (\mathbb{E}(W(a, b)))^2 \approx a^\beta \qquad (3)$$

where the exponent $\beta \in [-1, 3]$. In [19] a relationship between H and β for self-affine is found. For this reason, the definition of Hurst exponent is:

- $H = \frac{\beta+1}{2}$, for $\beta \in [-1, 1)$, if the signal is fGn,
- $H = \frac{\beta-1}{2}$, for $\beta \in [1, 3]$, if the signal is fBm.

The average wavelet coefficient method (AWC) proposed by Simonsen et al. [7] can be summarized in the following algorithm:

- Step 1: Wavelet transform is applied. Scaling and detail coefficients are obtained.
- Step 2: The averaged wavelet coefficients for a fixed scale j are calculated.
- Step 3: A log-log plot of detail coefficients versus scale a is done. The slope of the straight line is $\frac{1}{2} + H$.

2.2.2 The Modified Average Wavelet Coefficient Method

Improvements to AWC wavelet method described above can be obtained using more robust estimation methods in Step 2 [8].

We propose, given a series $x_1, \ldots, x_m$, to estimate the variance of detail coefficients (Eq. 3) with

- the *median absolut deviation* (MAD):

$$MAD(W(a, b)) = Med(|W(a, b) - Med(W(a, b))|) \qquad (4)$$

which is another measure of dispersion and corresponds to the median of the absolute deviations from the detail coefficient's median [20].

We call the AWC method, using this estimator, the *AWC-MAD* method.

3   Time Series

The data used in our study is described in this section. First we use synthetic series (Sect. 3.1) and in Sect. 3.2 rainfall series from different climatic regions of the Argentine Republic.

3.1   Synthetic Series

We generated 30 synthetic series for different Hurst exponent values (H = 0.2, 0.4, 0.6, 0.8), using MATLAB code wfbm.m. The programming sentence FBM = wfbm(H,L) returns a fractional Brownian motion signal of parameter H and length L, following the algorithm proposed by Abry and Sellan [1]. Each series has 32768 datapoints and Daubechies orthogonal wavelet Db10 (Fig. 1) is used [18].

In Figs. 2 and 3 examples of series fBm-type are shown.

Since fGn-type series represent the increments of fBm-type processes and both the signals are characterized by the same H value by definition, estimators designed for fGn-type series can be applied to fBm-type sequences after differentiation, and conversely estimators designed for fBm-type processes can be applied to fGn-type series after integration. fGn-type series are generated taking the first difference in fBm-type series, obtained with wfbm.m. Examples are shown in Figs. 4 and 5.

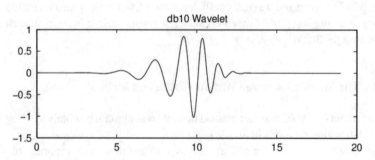

Fig. 1 Daubechies wavelet Db10

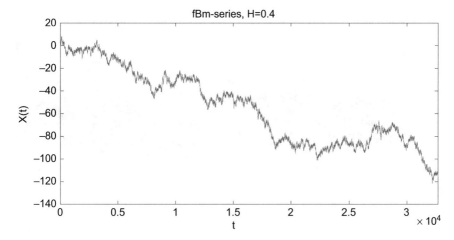

Fig. 2 Fractional Brownian motion fBm-type series with $H = 0.4$

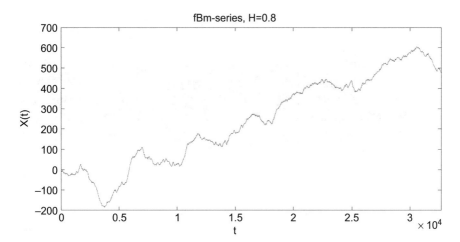

Fig. 3 Fractional Brownian motion fBm-type series with $H = 0.8$

3.2 Rainfall Series

The environment is a very important problem in the world today. Climate change is the process in which the weather is gradually changing because of pollution, among other issues.

Global warming means truly global warming: the atmosphere, the oceans, and the ground are all warming. As a result, ice is melting, seas are rising, storms are getting more severe, and floodings and droughts are getting worse. Since a few decades ago, researchers want to know how rainfall has changed during the seasons. The trend they found was clear – the rainy and drought seasons are increasing. As

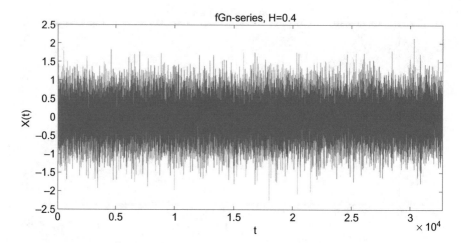

Fig. 4 Fractional Gaussian noise fGn-type series with $H = 0.4$

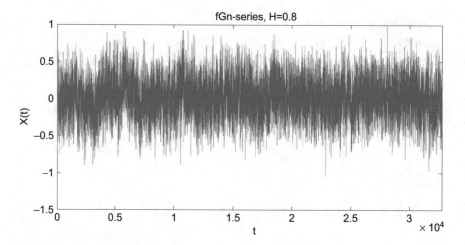

Fig. 5 Fractional Gaussian noise fGn-type series with $H = 0.8$

consequence, the study of rainfall series and their predictability is a very important task today. It was mentioned in Sect. 1 that rainfall series are an observable of a dynamical system that depends on different variables. For this reason, we not only study the time evolution of the series, but also we focus on their spatial behavior (latitude and longitude).

In the literature we found a unique report with rainfall series from La Pampa, Argentina [11]. Therefore, in a first stage we applied the methods described in Sect. 2 to these series. The second stage consists of analyzing different climatological regions of Argentina.

3.2.1 La Pampa's Rainfall Series

The report of Perez et al. [11] mentioned above is limited to some locations in the province of La Pampa, and the Hurst exponent was calculated using the R/S method for annual rainfall series. La Pampa is a province of Argentina, located in the center of the country, and it is dominated by mild climate. Precipitation is highly variable from year to year and generally decreases from east to west and from north to south. The area covers a strip of approximately 400 km in a latitudinal direction ($35°S$ to $38°30'S$) and 100 km in a longitudinal direction ($64°30'W$ to $63°30'W$), where the progress and decline of agriculture depend on rainfall. Mean annual precipitation ranges from 260 mm in the southwest to 820 mm in the northeast. In this work, we used time series corresponding to the same geographical locations and the same period of time (from 1960 to 2010), which can be seen in Fig. 6.

3.2.2 Argentina's Rainfall

To have an overview of the different climatic regions of Argentina, we have selected 34 cities, whose locations can be seen on the map (Fig. 7).

Argentina geographic regions are very dissimilar. There are plains and mountain ranges, woods and jungles, and arid, swampy, or clayish lands. Traversing the country along its latitude (3779 km) is a long way [21].

From the point of view of geography, Argentinian regions contemplated in this work can be divided into four different types of climate according to its most salient features:

• **Warm Climate:** It is found in the northeastern angle of Argentina.
• **Mild Climate:** The amount and distribution of rainfall determine two varieties of mild climate, to the east, pampas or humid weather with strong oceanic influence on the southeast coast of Buenos Aires, and a strip west, temperate transition occurs, to the arid climate.
• **Cold Climate:** The nival cold is characterized by permafrost, rainfall exceeding 800 mm, and westerly winds.
• **Arid Climate:** According to the altitude and latitude, this climate shows four varieties: high-mountain arid, sierras-and-fields arid, steppe arid, and cold arid.

Although Argentina is considered to have a temperate climate, it has very humid regions such as Misiones and regions very arid like San Juan. Figure 8 shows the rainfall series of the two provinces. One located in El Dorado, Misiones, and the other one in Campamento Filo de Sol, San Juan. When comparing them, there is a great variety in the range of rainfall and the dynamics.

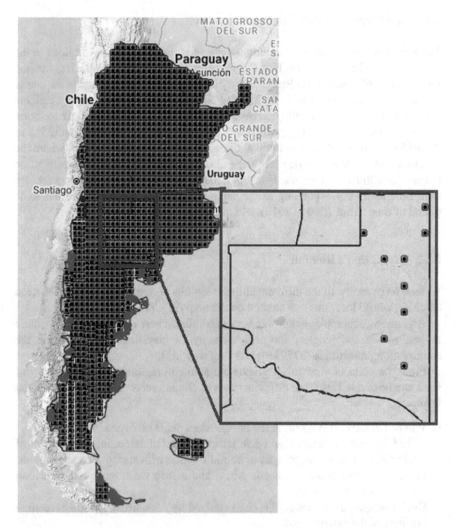

Fig. 6 Geographical locations from La Pampa, Argentina

4 Results

4.1 Synthetic Series

In the literature there are methods that are applicable to series of fBm-type and others that only can be applied to fGn series [2].

Tables 1 and 2 show the mean and standard deviation values computed for 30 synthetic series of type fBm and fGn, respectively. These results were obtained for each H value and after application of R/S, AWC, and AWC-MAD methods. It can

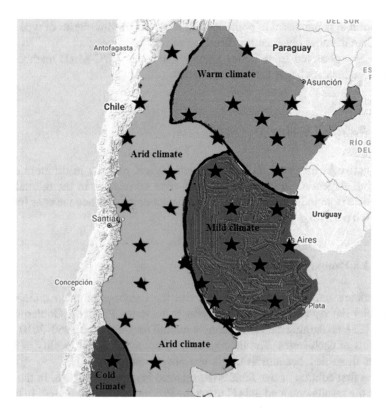

Fig. 7 Climatological distribution of República Argentina, and geographical locations

be observed from both tables that the AWC-MAD method reaches closer values to the true H than R/S and AWC. R/S is appropriate for nonstationary (fGn) series, so a previous analysis is necessary, while our method does not have this requirement.

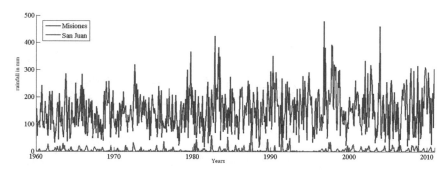

Fig. 8 Rainfall series, in millimeters, of Misiones and San Juan (1960 to 2010)

In addition, it is important to point out that H values for time series of type fGn are better than the ones obtained for fBm-type series.

It is also observed from Tables 1 and 2 that using AWC-MAD method, the H values obtained are better for fGn-type series.

4.2   Rainfall Series

An objective in this work is to study hydrological patterns in different climatic regions of Argentina in order to find long-term correlation in the rainfall series. Persistence is an indicator that present events not only influence the near future but will also have a long-term impact.

4.2.1   La Pampa's Rainfall Series

For compare our study with [11], we use the database of the Administración Provincial del Agua, Ministerio de Obras y Servicios Públicos, Gobierno de La Pampa [23], containing data about annual rainfall for the period 1960–2010, i.e., 84 datapoints of each series. The time series corresponding to the locality of Realico has been discarded, because its length was not as reported in [11].

In the first column of the Table 3 the selected locations are listed. In the second column the results obtained in [11] using the R/S method are presented. In the third column, R/S* are the results of the R/S method applied to the database in [23]. Finally, the last column corresponds to H values which were obtained using the AWC-MAD method.

Table 3 shows that the difference between the values presented in [11] and the ones we obtained is lower than 0.3. In all cases time series are persistent. Furthermore, the values of H that were obtained applying the wavelet method are lower than those with R/S methods. However, it is important to point out that

Table 1 Hurst exponent values for fBm-type series

H	R/S	AWC	AWC-MAD
0.2	0.9847 ± 0.0105	0.1698 ± 0.0635	0.1803 ± 0.0658
0.4	1.0048 ± 0.0153	0.3692 ± 0.0564	0.3826 ± 0.0555
0.6	1.0071 ± 0.0119	0.5505 ± 0.0562	0.5647 ± 0.0566
0.8	1.0074 ± 0.0063	0.7042 ± 0.0529	0.7189 ± 0.0550

Table 2 Hurst exponent values for fGn-type series

H	R/S	AWC	AWC-MAD
0.2	0.3197 ± 0.0080	0.1963 ± 0.0252	0.2081 ± 0.0226
0.4	0.4706 ± 0.0108	0.3916 ± 0.0257	0.4038 ± 0.0306
0.6	0.6329 ± 0.0156	0.5846 ± 0.0318	0.5939 ± 0.0339
0.8	0.7880 ± 0.0121	0.7774 ± 0.0352	0.7883 ± 0.0424

Table 3 Values of the Hurst exponent for rainfall series of La Pampa

Locality	R/S	R/S*	AWC-MAD
Alpachiri	0.86	0.853	0.496
BLarroulde	0.91	0.874	0.775
Bernasconi	0.82	0.819	0.614
Doblas	0.82	0.874	0.738
ECastex	0.96	0.835	0.712
GPico	0.96	0.896	0.715
Guatrache	0.97	0.787	0.780
Lonquimay	0.93	0.917	0.744
SRosa	0.77	0.838	0.310

Table 4 Hurst exponent values for warm climate

Province	Locality	Lat. ; Long.	AWC-MAD
Formosa	Com. La Rinconada	23°5′ ; 61°5′	0.2668
	Formosa	26°5′ ; 58°5′	0.3023
Santiago del Estero	Alberdi	26°5′ ; 62°5′	0.4500
Chaco	Santa Sylvina	27°5′ ; 61°5′	0.5365
Tucumán	La Cocha	27°5′ ; 65°5′	0.4320
Corrientes	Gral. Alvear	28°5′ ; 56°5′	0.5602
Santa Fé	Puerto Ocampo	28°5′ ; 59°5′	0.4810
Entre Ríos	San Gustavo	30°5′ ; 59°5′	0.4305

Alpachiri and Santa Rosa series would be anti-persistent based on the wavelet method results.

4.2.2 Argentina Rainfall Series

Data sets were collected from Base de Datos Climáticos de la República Argentina [22] corresponding to monthly rainfall ranging from 1960 to 2010. We take monthly data because for the daily data, there are many zeros in the real series not simulated. As described in Sect. 3.2.2, Argentina has a wide variety of climates, from wet to dry, of tropical heat or cold snow, through different types of mild climates. For this reason we choose series from different regions, grouping them according to the type of climate.

As it was mentioned before, Hurst exponent is a measure of the degree of dependency of the rainfall time series. In Table 4, we present warm weather series, indicating the geopolitical location: province to which they belong, locality and latitude and longitude, and the values of H estimated with AWC-MAD method, in the last column.

In particular, we observe that they are anti-persistent series, except for Chaco and Corrientes.

Table 5 Hurst exponent values for mild climate

Province	Locality	Lat. ; Long.	AWC-MAD
Córdoba	Leones	32°5′ ; 62°5′	0.2063
	Obispo Trejo	30°5′ ; 63°5′	0.2673
La Pampa	Colonia Santa Teresa	37°5′ ; 63°5′	0.5345
	Santa Rosa	36°5′ ; 64°5′	0.6277
Entre Ríos	San Gustavo	30°5′ ; 59°5′	0.4305
	Gdor. Mansilla	32°5′ ; 59°5′	0.2978
CABA	CABA	34°5′ ; 58°5′	0.3317
Buenos Aires	Blaquier	34°5′ ; 62°5′	0.5494
	Junín	35°5′ ; 60°5′	0.5391
	Tandil	37°5′ ; 59°5′	0.3784
	Bahía Blanca	38°5′ ; 62°5′	0.5654

In the case of the mild climate, we have selected 11 series (see Table 5), whose locations follow the previous table. Due to its climatic characteristics and flat relief (or pampas plain), this region is important for agricultural or livestock development, and it is also where the largest population is concentrated.

The values of this region range from the anti-persistent to the persistent as we move southwest, except for the mountain town of Tandil.

The arid climate is the one that covers the greatest territory and is subdivided into four varieties, depending on the geography. Table 6 shows 14 time series distributed throughout the region.

Table 6 Hurst exponent values for arid climate

Province	Locality	Lat. ; Long.	AWC-MAD
Jujuy	Susques	23°5′ ; 66°5′	0.4314
Catamarca	Cerro El Cóndor	26°5′ ; 68°5′	0.4419
San Juan	Campamento Filo del Sol	28°5′ ; 69°5′	0.7175
La Rioja	Tama	30°5′ ; 66°5′	0.4945
Mendoza	Puente del Inca	32°5′ ; 69°5′	0.6757
	San Rafael	34°5′ ; 68°5′	0.4225
	Malargüe	36°5′ ; 68°5′	0.6589
Santiago del Estero	Mailín	28°5′ ; 63°5′	0.2721
San Luis	La Pampa Grande	32°5′ ; 66°5′	0.4695
	Arizona	35°5′ ; 65°5′	0.3455
La Pampa	Gdor. Duval	38°5′ ; 66°5′	0.5471
Buenos Aires	Viedma	40°5′ ; 62°5′	0.5324
Neuquén	Añelo	38°5′ ; 69°5′	0.5935
Río Negro	Sierra Colorada	40°5′ ; 67°5′	0.6374

Table 7 Hurst exponent
values for cold climate

Province	Locality	Lat. ; Long.	AWC-MAD
Neuquén	Achico	$40°5'$; $70°5'$	0.5018

The values of AWC-MAD greater than 0.5 are observed in the series that correspond to the arid mountain climate (long. $69.5°$) and Sierra Colorada, near the cold climate.

Finally, we wanted to consider the four types of weather in this study, but in the case of cold weather, we had only one series, since the database [22] has no information below the $40°S$ parallel, neither from the island territories in the South Atlantic. In Table 7 the near random value of H is shown.

5   Conclusions

We present a surpassing proposal to the AWC method, called AWC-MAD method. Both techniques use the wavelet transform considering several time scales, so they have a predictive power related to the self-affinity of the series.

Results obtained for synthetic series show that AWC-MAD method has a better performance than the R/S, the most used method to compute Hurst exponent, still when the nature of the series is known. This fact encouraged us to use AWC-MAD to compute Hurst exponent in rainfall series.

The prediction of meteorological variables is a subject of a great deal of research, given the concern about climate change. However, it is recent the emergence of studies that consider the dynamics of these systems and fractality to characterize the self-affinity of the time series.

In this paper we deal with the characterization of rainfall series in Argentina. First, we compare AWC-MAD with the unique result found in the literature in which the authors used R/S in annual rainfall series from La Pampa, Argentina. Then, we compute H of monthly series from different climatological regions in Argentina. As future work, we want to test the AWC-MAD method with time series of different nature, such as financial time series. With respect to the rainfall series, it is suggested that correlation with other meteorological variables should give a more complete characterization of their dynamics. Further analysis with larger databases is also planned.

References

1. Abry, P., Sellan, F.: The wavelet-based synthesis for the fractional Brownian motion. Proposed by F. Sellan and Y. Meyer: remarks and fast implementation. Appl. Comput. Harmon. Anal. **3**, 377–383 (1996)

2. Serinaldi, F.: Use and misuse of some Hurst parameter estimators appied to stationary and non-stationary financial time series. Physica A Stat. Mech. Appl. **389**(14), 2770–2781 (2010)
3. López-Lambraño, A.A., Fuentes, C., López-Ramos, A.A., Mata Ramírez, J., Lopez-Lambraño, M.: Spatial and temporal Hurst exponent variability of rainfall series based on the climatological distribution in a semiarid region in Mexico. Atmósfera **31**(3), 199–219 (2018)
4. Bariviera, A., Martín, M.T., Plastino, A., Vampa, V.: LIBOR troubles: Anomalous movements detection based on Maximum Entropy. Phys. A **449**, 401–407 (2016)
5. Hurst, H.E.: Long-term storage capacity of reservoirs. Trans. Am. Soc. Civ. Eng. **116**, 770–808 (1951)
6. Mandelbrot, B.B., Van Ness, J.W.: Fractional Brownian motions, fractional noises and applications. SIAM Rev. **10**(4), 422–437 (1968)
7. Simonsen, I., Hansen, A., Nes O.: Determination of the Hurst exponent by use of wavelet transforms. Phys. Rev. E Am. Phys. Soc. **58**(3), 2779 (1998)
8. Arouxet, M.B., Pastor, V.: Estudio del exponente de Hurst, Mecánica Computacional, 35, 2501–2508, Proceeding of Congreso sobre Métodos Numéricos y sus Aplicaciones (2017)
9. Arouxet, M.B., Pastor, V.: Modificación al método wavelet y su aplicación en series de precipitaciones, LXVII Reunión Anual de Comunicaciones Científicas-UMA (2018)
10. Salomão, L., Campanha, J., Gupta, H.: Rescaled range analysis of pluviometric records in São Paulo State, Brazil. Theor. Appl. Climatol. **95**, 83–89 (2009)
11. Pérez, S., Sierra, E., Massobrio, M., Momo, F.: Análisis fractal de la precipitación anual en el este de la provincia de La Pampa, Revista de Climatología (2009)
12. Cannon, M.J., Percival, D.B., Caccia, D.C., Raymond, G.M., Bassingthwaighte, J.B.: Evaluating scaled windowed variance methods for estimating the Hurst coefficient of time series. Phys. A **241**, 606–626 (1997)
13. Hurst, H.E.: Methods of using long-term storage in reservoirs. Proc. Inst. Civ. Eng. **1**, 519–543 (1956)
14. Zunino, L., Tabak, B.M., Figliola, A., Perez, D.G., Garavaglia, M., Rosso, O.A.: A multifractal approach for stock marketing inefficiency. Phys. A **387**, 6558–6566 (2009)
15. Chui, C.K.: An Introduction to Wavelet Analysis. Academics (1992)
16. Walnut, D.: An Introduction to Wavelet Analysis, Applied and Numerical Harmonic Analysis Series. Birkhuser Eds., Boston (2002)
17. Mallat, S.G.: A Wavelet Tour of Signal Processing: The Sparse Way. Academic, Elsevier (2009)
18. Daubechies, I.: Ten Lectures. Academics (1988)
19. Malamud, B., Turcote, D.: Self-affine time series: measures of weak and strong persistence. J. Stat. Plann. Inference **80**, 173–196 (1999)
20. Maronna, R.A., Martín, R.D., Yohai, V.J.: Robust Statistics: Theory and Methods. Wiley (2006)
21. Farber, M.E., Raizboim, I.N.: El Sur del Sur: Argentina, el país, su cultura y su gente. https://surdelsur.com/
22. http://3cn.cima.fcen.uba.ar
23. https://apa.lapampa.gob.ar/metereologias/19-meteorologia/42-datos-historicos-de-lluvia.html

Application of Wavelet Transform to Damage Detection in Brittle Materials via Energy and Entropy Evaluation of Acoustic Emission Signals

Juan P. Muszkats, Miguel E. Zitto, Miryam Sassano, and Rosa Piotrkowski

Abstract Acoustic emission (AE) hits from uniaxial compression tests of andesite rock samples were processed with the continuous wavelet transform (CWT). The quest for frequency bands with minimum entropy values arrived at 150 and 250 kHz as those related to macro-fracture mechanisms. A preprocessing algorithm was developed in order to attenuate the influence of reflected signals at the inner interfaces of the material. It is based on the detection of abrupt phase changes of the CWT coefficients. Entropy calculations performed with the hits already processed permitted a reliable study of the AE entropy evolution in the relevant frequency bands and its relationship with the corresponding cumulative AE energy evolution.

1 Introduction

Compressive failure in brittle materials, like rocks submitted to load, consists of an alternation of cracking processes: micro-cracks initiation at certain preexistent flaws followed by micro-cracks coalescence into macro-cracks, macro-cracks growth and branching into new micro-cracks, fragmentation, and final collapse. Thus, tracking of macro-crack initiation and growth seems to be an appealing method for studying

J. P. Muszkats (✉)
Facultad de Ingeniería, Universidad de Buenos Aires, Buenos Aires, Argentina

Universidad Nacional del Noroeste de la Provincia de Buenos Aires, Buenos Aires, Argentina
e-mail: jpmuszkats@comunidad.unnoba.edu.ar

M. E. Zitto
Facultad de Ingeniería, Universidad de Buenos Aires, Buenos Aires, Argentina

M. Sassano
Universidad Nacional de Tres de Febrero, Buenos Aires, Argentina

Facultad de Ingeniería, Universidad de Buenos Aires, Buenos Aires, Argentina

R. Piotrkowski
Escuela de Ciencia y Tecnología, Universidad Nacional de San Martín, Buenos Aires, Argentina

Facultad de Ingeniería, Universidad de Buenos Aires, Buenos Aires, Argentina

damage evolution in brittle materials [1–4]. The energy applied (load) to the rocky material is transferred and stored as strain-stress energy in different locations. In those locations where thresholds are suddenly surpassed, energy is dissipated in fracture energy or surface energy. As in any dissipative process, the causes of damage can be described within the frame of the irreversible thermodynamic framework. In this sense, damage parameters, such as entropy generation, account for degradation and loss of structural integrity [5–8].

Acoustic emission (AE) consists of elastic waves generated in the interior of materials. These waves are induced by a rapid change in the stress-strain condition around a given point. In the case of brittle materials, nucleation, advance, opening, and closure of fractures are the main sources for these waves. AE propagates undergoing attenuation and multiple reflections, especially in heterogeneous materials. Eventually, the waves reach the surface, where they can be detected by piezoelectric sensors that transform them into electrical signals, i.e., the AE signals, which are then processed for further analysis. AE signals are hits of very low amplitude (about $10\,\mu V$) and high frequency (ranging from $1\,kHz$ to $1\,MHz$), so they have to be immediately amplified. AE equipment stores the waveforms and calculates characteristic parameters such as energy, root mean square (RMS) value, amplitude, rise time, event duration, etc. [9].

AE is ultimately generated by the rupture of atomic bonds. It involves different spatial and temporal scales ranging from microscopic events to seismic faults. Because of this, seismic information and AE in rocks are complementary, both in their applications and in their theoretical basis. Furthermore, AE has been established decades ago as a well-suited tool to evaluate the dynamic state of bulk and surface defects [10]. This can be accomplished by analyzing elastic waves emitted during micro-fracture processes. It is nowadays used in material science and engineering research, including work reported by the authors [2, 3, 11, 12]. Because AE is typically a nonstationary process, wavelet transform (WT) is an appropriate tool for these studies [13].

In previous work [2] the complex Morlet continuous wavelet transform (CWT) was applied to AE signals from dynamic tests conducted on a reinforced concrete slab with a shaking table. The frequency band corresponding to the fracture of concrete was identified by comparing the scale position of maximum CWT values with the response acceleration obtained in seismic simulations. The AE signals were reconstructed in this scale (frequency) band, and cumulative acoustic emission energy (CAE) was calculated. The resulting CAE was compared with the cumulative dissipated energy (CDE) of the tested structure, an accepted parameter for characterizing the mechanical damage in structures: a good agreement was found between the normalized histories of CAE and CDE. Thus, the particular scale (frequency) in which AE signals were reconstructed could be attributed to the fracture of concrete.

In more recent work [14, 15], CWT was applied to AE signals resulting from uniaxial compression tests of andesite rock samples up to breakage. AE signals were filtered into different frequency bands with the CWT. Some of these bands were identified as characteristic frequencies of the fracture process, in accord with a physical model of seismic focus that describes the advance and propagation of waves during the fracture of brittle materials [4]. Precise results were achieved considering the nonstationary nature of the involved physical processes. The wavelet energy b-value, a variant of the Gutenberg-Richter law that rules in geophysics [3, 16], was successfully applied to trace the inception of dangerous cracks.

In the present paper, which continues [14, 15], we intend to go further into the detection and evaluation of macro-fracture in rocks. To this end entropy and phase studies are introduced in our processing of AE signals.

Irreversible thermodynamic processes cause degradation of materials, and damage is a phenomenon with increasing disorder. The energy dissipated in damage (fracture) results in entropy increase, according to the second law of thermodynamics. Thus, it is important to investigate the damage-entropy relationship while loading material [17]. The premise of Gibbs physical entropy as a limit of the mathematical Shannon entropy is demonstrated in [18] and applied in empirical work in complex systems in other fields [19]. In recent works, Shannon entropy is successfully applied to AE signals for detecting damage in different rocky materials [20, 21]. In the present work, the wavelet Shannon entropy is applied twofold to the same signals as those analyzed in [15]. Firstly, to detect specific signals coming from macro-crack nucleation and advance. Secondly, to follow precisely and concisely the macro-fracturing evolution of andesite rock under load. Another novelty from previous work is that the signals under study were preprocessed with a specifically developed technique. This technique focuses on the phase of complex wavelet coefficients and allowed us to reduce the distortion caused by wave reflections on the detected signals.

2   Experiment

Four cylindrical andesite rock samples were tested as described in [22]. These specimens from Cerro Blanco, San Juan, Argentina, were 75 mm in diameter and 150 mm in length. As illustrated in Fig. 1, the rock samples were subjected to uniaxial compression up to rupture. The equipment consisted of a CGTS machine with a 100 tons capacity of servo-hydraulic type and a closed loop. The actuator displacement speed was 0.12 mm/min.

AE was monitored with three piezoelectric sensors. The present work focuses on the results gathered by the broadband sensor (100–1000 kHz) in one of the samples. The sampling frequency was 1 MHz, the AE system (also shown in Fig. 1) was a PCI-2 two-channel PAC plate, and the commercial software AEWIN was used for the initial determination of classical AE parameters. This experiment resulted in a collection of about 75,000 AE hits, saved in individual files. The analysis of such

Fig. 1 The experimental setup and the AE system

a big quantity of information proved to be time-consuming but manageable for a
standard computer.

3   Mathematical Resources

This section contains the description and definition of techniques that we have
been consistently using throughout our previous work [14, 15], as well as entropy
definitions and the algorithm developed in the present work for preprocessing data.

3.1   *Continuous Wavelet Transform*

The continuous wavelet transform (CWT) is defined in [23] by means of a
continuous wavelet function $\psi(t)$. This function must verify an exponential decay
and also that $\int_{\mathbb{R}} \psi = 0$. Given a function $f \in L^2(\mathbb{R})$, its CWT is defined as

$$c(j, k) = \frac{1}{\sqrt{|j|}} \int_{-\infty}^{\infty} f(t) \, \overline{\psi\left(\frac{t - k}{j}\right)} \, dt \tag{1}$$

if $j \neq 0$, while $c(0, k) = 0$. In each CWT coefficient $c(j, k)$, the value of j
indicates a scale (and therefore a frequency), while k denotes time displacement.

In the present work, we used the Morlet wavelet defined by

$$\psi(t) = \pi^{-\frac{1}{4}} \cdot e^{6it} \cdot e^{-\frac{t^2}{2}}$$ (2)

Being continuously defined for every j and k, practical implementation arises numerical issues. Approximation algorithms and details can be found in [24].

In order to illustrate the CWT when applied to AE, the plot of a typical hit is shown in the upper part of Fig. 2. Its corresponding scalogram in the lower part of Fig. 2 illustrates the distribution in time and frequency of the wavelet energy density $|c(j, k)|^2$, with value increasing from darker to lighter tones.

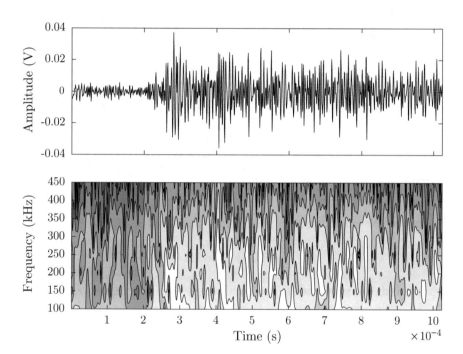

Fig. 2 A typical hit. Upper: waveform. Lower: scalogram (increasing from darker to lighter tones)

The simplest way to perform bandpass filtering with the CWT consists in reconstructing the signal only with the CWT coefficients of the desired frequencies. Figure 3 illustrates this procedure applied to the same previous hit, filtered at 250 kHz. The bandpass frequency is clear in the corresponding scalogram, shown in Fig. 3.

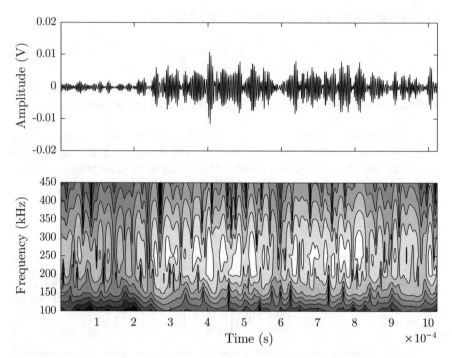

Fig. 3 Same hit as in Fig. 2 filtered around 250 kHz. Upper: waveform. Lower: scalogram (increasing from darker to lighter tones)

The acoustic emission energy (AEE) of a hit is defined in [3] as

$$AEE = \sum x_i^2 \, \Delta t \tag{3}$$

where x is the reconstructed signal after filtering and Δt is the sample rate.

3.2 Wavelet Entropy

The Shannon entropy can be adapted as in [11] to obtain a measure of the intrinsic order in a signal by means of its CWT coefficients. When CWT is numerically implemented, only a finite quantity of the c_{jk} coefficients defined by (1) can be obtained. Thus, the wavelet power (WP) corresponding to the jth scale is defined by

$$WP_j = \sum_{k=1}^{N} \left| c_{jk} \right|^2 \tag{4}$$

The fraction of wavelet power corresponding to time k is expressed by the p_{jk} coefficients:

$$p_{jk} = \frac{|c_{jk}|^2}{W P_j} \qquad (5)$$

With these elements, wavelet entropy of the jth band is formally equal to the Shannon entropy:

$$S_j = -\sum_{k=1}^{N} p_{jk} \log p_{jk} \qquad (6)$$

Figure 4 shows for the hit previously considered the values of entropy calculated for the different scales (frequencies). Coalescence of micro-cracks into macro-cracks, which implies the transition from less to more organized structures, is expected to express in frequencies with lower entropy. Therefore, those frequencies for which relative minima of entropy occur are of special interest.

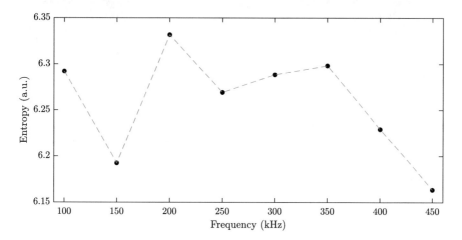

Fig. 4 The same hit as in Fig. 2 reaches relative minima of entropy for the 150 and 250 kHz bands

3.3 Preprocessing

Each detected AE hit is the complex consequence of a large number of random events [25]. In particular, reflections at the multiple interfaces present in natural rocks constitute a special problem. If a second mechanical pulse reaches the detector while the first one is still operating, the apparent duration, the waveform, and the entropy of this composite signal are affected. To reduce the effect of reflections, a special algorithm was developed. Given that the Morlet wavelet coefficients are complex numbers, it is expected to find their phase varying cyclically when analyzing a perfectly periodic signal. When reflections occur, they imply the sudden overlapping of vibrations, with the consequent phase changes. These phase variations are detected by studying the deviations from the expected cyclical plot. Figure 5 shows, for the same hit previously studied, the phase plot of the CWT coefficients corresponding to the 250 kHz scale. The square denotes the instant when the amplitude surpasses a threshold for the first time. Diamonds mark the instants when sudden phase changes occur. Those diamonds closest to the square delimit the relevant part of the hit, preserved for ulterior entropy calculations.

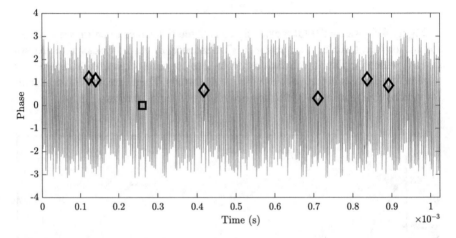

Fig. 5 Same hit as in Fig. 2 filtered around 250 kHz: phase of wavelet coefficients versus time. Square: amplitude surpasses a threshold for the first time. Diamonds: sudden phase change

Thus, the most relevant (and least distorted) part of each hit is selected, and this analysis is performed separately for each scale (see Fig. 6). For our present work, it was only after the described preprocessing that entropy for every hit and scale was calculated.

4   Results and Discussion

Despite being continuously defined in (1), the CWT can only be calculated for a discrete set of values j and k. Moreover, the Heisenberg uncertainty principle imposes an unavoidable trade-off between time and frequency localization. Therefore, the choice of the scales for which the CWT is calculated must rely upon a trustworthy frequency resolution. In the present work, several features were taken into account: the range of the broadband sensor (100–1000 kHz), the sampling frequency (1 MHz), and the duration of the shortest hits (estimated to be

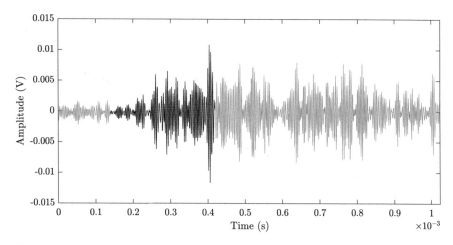

Fig. 6 Preprocessed hit. Ligher plot: original filtered hit. Darker plot: preserved section of the hit

about $10\,\mu s$). Under these premises, the width of frequency bands was chosen to be $50\,kHz$. Figures 2 and 3 illustrate this choice.

In previous work [14, 15] the detection of characteristic frequencies relied upon energy considerations. Frequency bands near 150 and $250\,kHz$ were chosen to be those most representative of the fracture mechanism. Figure 7 plots the cumulative AEE calculated according to (3). The signal with its whole frequency content is plotted as a dashed line, while the signal reconstructed at $150\,kHz$ is plotted in black, and the signal reconstructed at $250\,kHz$ is plotted in gray. Both filtered signals and the original signal show similar energy evolution and jumps. The 150 and the $250\,kHz$ reconstructed signals also show a very similar energy level throughout the whole experiment.

The entropy concept provides a complementary tool that proved to be consistent with previous results. For the present purposes, entropy is considered as a measure of disorder in a signal. Therefore, lower entropy values suggest the occurrence of more organized phenomena. After the preprocessing already described, every single hit was analyzed as in Fig. 4. Frequencies with relative minimum entropy were detected according to a threshold criterion (eventually, a single hit might have several relative minima).

Figure 8 is a histogram which accounts for the density of hits that present a minimum entropy at a given frequency band. Table 1 gathers basic statistical information about entropy and frequency after analyzing the whole set of hits. It also shows the average time duration of the preprocessed hits, discriminated by frequency band. This duration showed to be consistently smaller as frequency increases.

According to Table 1, 150 and $250\,kHz$ frequencies turned out to be those for which entropy reaches a relative minimum more often. It is also important to notice

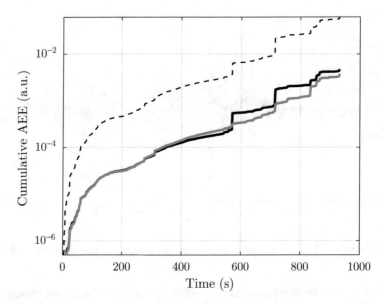

Fig. 7 Cumulative AEE along the test. Dashed: whole frequency content. Black: 150 kHz. Gray: 250 kHz

Table 1 Preprocessed hits statistics

Frequency (kHz)	Number of entropy minima	μ (entropy mean value)	σ (entropy standard deviation)	Average time duration (miliseconds)
100	891	5.7	0.5	0.65
150	26847	4.9	0.7	0.37
200	16612	4.8	0.8	0.37
250	20988	4.6	0.8	0.29
300	9795	4.5	0.8	0.25
350	9291	3.7	1	0.12

that these minima were mostly reached by the end of the experiment, as shown in Fig. 8. Therefore, these hits are of special interest and most likely to express a characteristic frequency of the fracturing process. Figure 9 shows the entropy and its time evolution for those hits which attain a minimum at 150 kHz (black line) and those which attain a minimum at 250 kHz (gray line). In order to make it easier to read the plot, a moving mean of 1000 points was applied to the results. The entropy value for the 250 kHz proved to be consistently smaller throughout the whole experiment. This would imply that the 250 kHz frequency band is more strongly related to the advance of macro-cracks, a more organized phenomenon.

The AEE defined in (3) can be calculated for reconstructed hits after bandpass filtering. It is of particular interest to trace the time evolution of this energy: therefore the cumulative AEE for each frequency band was calculated. Upper

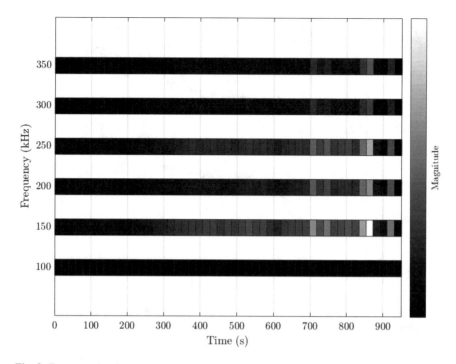

Fig. 8 Frequency bands and density of hits with minimum entropy

Fig. 10 shows the cumulative AEE for the 150 kHz band, while lower Fig. 10 shows the cumulative AEE for the 250 kHz band (both in thin lines). These graphs also show the time evolution of the corresponding entropy already shown in Fig. 9. It can be appreciated that entropy evolution changes significantly, even shows relative minima, before cumulative energy jumps. That is, it anticipates the expression of dangerous damage.

Fracture in rocks is an interplay of macro-fracture, nucleation, advance and branching into micro-cracks, followed by further coalescence of micro-cracks into macro-cracks. The results displayed in Figs. 7, 8, 9, and 10 suggest that the main features of the whole fracturing process in andesite can be followed by the evolution of wavelet energy and wavelet entropy in the selected bands. Moreover, the 250 kHz band, due to its lower entropy values, seems to be connected to nucleation and advancement of macro-cracks.

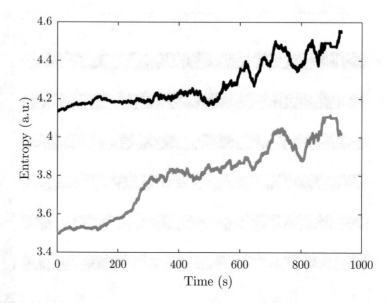

Fig. 9 Entropy of preprocessed hits at 150 kHz (black) and 250 kHz (gray). A moving mean of 1000 points was applied

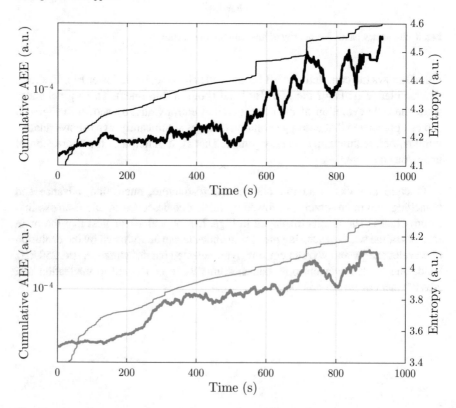

Fig. 10 AEE (thin) and entropy (thick). Upper: at 150 kHz. Lower: at 250 kHz

5 Conclusions

The minimum entropy criterion applied to AE hits resulted in a reliable tool for the detection of relevant frequency bands. It is sensibly improved by the preprocessing of the signals, which reduces the distorting effects due to inner reflections in the studied material.

Comparison with previous work and other criteria [14, 15] reinforces the conclusion that the 150 and 250 kHz are characteristic frequencies of the andesite rock, related to macro-fracture mechanisms.

Those hits most related to each of these bands were specifically studied. Their entropy evolved in accordance with the cumulative AEE. Major changes in AEE are anticipated and accompanied by abrupt oscillations in entropy. These sudden changes, both in AEE and entropy, indicate an abrupt reorganization of the material rheological state.

Acknowledgments The present work received financial support and is part of the Program UBACyT 20020160100038BA.

References

1. Huq, F., Liu, J., Tonge, A.L., Graham-Brady, L.: A micromechanics based model to predict micro-crack coalescence in brittle materials under dynamic compression. Eng. Fract. Mech. (2019). https://doi.org/10.1016/J.ENGFRACMECH.2019.106515
2. Zitto, M.E., Piotrkowski, R., Gallego, A., Sagasta, F., Benavent-Climent, A.: Damage assessed by wavelet scale bands and b-value in dynamical tests of a reinforced concrete slab monitored with acoustic emission. Mech. Syst. Signal Process. (2015). https://doi.org/10.1016/J.YMSSP.2015.02.006
3. Sagasta, F., Zitto, M.E., Piotrkowski, R., Benavent-Climent, A., Suarez, E., Gallego, A.: Acoustic emission energy b-value for local damage evaluation in reinforced concrete structures subjected to seismic loadings. Mech. Syst. Signal Process. (2018). https://doi.org/10.1016/j.ymssp.2017.09.022
4. Filipussi, D., Piotrkowski, R., Ruzzante, J.: Characterization of a crack by the acoustic emission signal generated during propagation. Procedia Mater. Sci. (2012). https://doi.org/10.1016/j.mspro.2012.06.036
5. Amiri, M., Modarres, M.: An entropy-based damage characterization. Entropy (2014). https://doi.org/10.3390/e16126434
6. Imanian, A., Modarres, M.: A thermodynamic entropy-based damage assessment with applications to prognostics and health management. Struct. Health Monit. (2018). https://doi.org/10.1177/1475921716689561
7. Vaughn, N., Kononov, A., Moore, B., Rougier, E., Viswanathan, H., Hunter, A.: Statistically informed upscaling of damage evolution in brittle materials. Theor. Appl. Fract. Mech. (2019). https://doi.org/10.1016/J.TAFMEC.2019.04.012
8. Kang, Y., Liu, H., Aziz M., Kassim, K.A.: A wavelet transform method for studying the energy distribution characteristics of microseismicities associated rock failure. J. Traffic Transp. Eng. (Engl. Ed.) (2019). https://doi.org/10.1016/J.JTTE.2018.03.007

9. Grosse, C., Ohtsu, M. (eds.): Acoustic Emission Testing. Springer, Heidelberg (2008)
10. Ono, K. : Acoustic Emission. In: Rossing T.D. (ed.) Springer Handbook of Acoustics. Springer Handbooks. Springer, New York (2014)
11. Piotrkowski, R., Castro, E., Gallego, A.: Wavelet power, entropy and bispectrum applied to AE signals for damage identification and evaluation of corroded galvanized steel. Mech. Syst. Signal Process. (2009). https://doi.org/10.1016/j.ymssp.2008.05.006
12. Piotrkowski, R., Gallego, A., Castro, E., García-Hernandez, M.T., Ruzzante, J.E.: Ti and Cr nitride coating/steel adherence assessed by acoustic emission wavelet analysis. NDT & E Int. (2005). https://doi.org/10.1016/J.NDTEINT.2004.09.002
13. Meyer, Y., Ryan, R.: Wavelets: algorithms & applications. Society for Industrial and Applied Mathematics, Philadelphia (1993)
14. Filipussi, D., Muszkats, J., Sassano, M., Zitto, M., Piotrkowski, R.: Fractura de roca andesita y análisis espectral de señales de emisión acústica. Tecnura. **23**(61), 45–56 (2019)
15. Muszkats, J.P., Filipussi, D., Zitto, M.E., Sassano, M., Piotrkowski, R.: Detection of fracture regimes in andesite rock via the energy evolution of acoustic emission signals in relevant frequency bands. In: Ceballos, L., Gariboldi, C., Roccia, B. (eds.) VII Congreso de Matemática Aplicada, Computacional e Industrial, pp. 489–492. ASAMACI, Río Cuarto, Córdoba (2019)
16. Rao, M.V.M.S., Prasanna Lakshmi, K.J.: Analysis of b-value and improved b-value of acoustic emissions accompanying rock fracture. Curr. Sci. (2005). https://doi.org/10.2307/24110936
17. Mahmoudi, A., Mohammadi, B.: On the evaluation of damage-entropy model in cross-ply laminated composites. Eng. Fract. Mech. (2019). https://doi.org/10.1016/J.ENGFRACMECH.2019.106626
18. Truffet, L.: Shannon entropy reinterpreted. Rep. Math. Phys. (2017). https://doi.org/10.1016/S0034-4877(18)30050-8
19. Dobovišek, A., Markovič, R., Brumen, M., Fajmut, A.: The maximum entropy production and maximum Shannon information entropy in enzyme kinetics. Phys. A Stat. Mech. Appl. (2018). https://doi.org/10.1016/J.PHYSA.2017.12.111
20. Chai, M., Zhang, Z. Duan, Q.: A new qualitative acoustic emission parameter based on Shannon's entropy for damage monitoring. Mech. Syst. Signal Process. (2018). https://doi.org/10.1016/J.YMSSP.2017.08.007
21. Bressan, G., Barnaba, C., Gentili, S., Rossi, G.: Information entropy of earthquake populations in northeastern Italy and western Slovenia. Phys. Earth Planet. Inter. (2000). https://doi.org/10.1016/J.PEPI.2017.08.001
22. Filipussi, D.A., Guzmán, C.A., Xargay, H.D., Hucailuk, C., Torres, D.N.: Study of acoustic emission in a compression test of andesite rock. Procedia Mater. Sci. (2015). https://doi.org/10.1016/J.MSPRO.2015.04.037
23. Boggess, A., Narcowich, F.J.: A First Course in Wavelets with Fourier Analysis, 2nd edn. Wiley, Hoboken (2009)
24. Torrence C., Compo G.P.: A practical guide to wavelet analysis. Bull. Am. Meteorol. Soc. (1998). https://doi.org/10.1175/1520-0477(1998)079<0061:APGTWA>2.0.CO;2
25. López Pumarega, M.I., Armeite, M., Ruzzante, J.E., Piotrkowski, R.: Relation between amplitude and duration of acoustic emission signals. Rev. Quant. Nondestruct. Eval. **22**, 1431–1438 (2003)

Printed in the United States
by Baker & Taylor Publisher Services